本书受到河北省高等教育教学改革研究与实践项目（
政合作河北衡水湖湿地保护与恢复项目、衡水学院河北
目、河北省中央引导地方科技发展资金项目（236Z7602G、226Z4202G）资助

水 科 学

可持续发展教育（ESD）活动指南

Water Science: Education for Sustainable
Development (ESD) Guidebook

·汉英对照·

武大勇　　［德］圭窦·库克曼斯特　　吴军梅

主编

燕山大学出版社
·秦皇岛·

图书在版编目（CIP）数据

水科学：汉英对照 / 武大勇，（德）圭窦·库克曼斯特，吴军梅主编. —秦皇岛：燕山大学出版社，2024.9

（可持续发展教育（ESD）活动指南）

ISBN 978-7-5761-0656-5

Ⅰ. ①水… Ⅱ. ①武… ②圭… Ⅲ. ①水文学－青少年读物－汉、英 Ⅳ. ①P33-49

中国国家版本馆 CIP 数据核字（2024）第 063642 号

水科学
SHUIKEXUE

武大勇 [德]圭窦·库克曼斯特 吴军梅 主编

出 版 人：陈　玉			
责任编辑：张　蕊		策划编辑：张　蕊	
责任印制：吴　波		封面设计：刘馨泽	
出版发行：燕山大学出版社 YANSHAN UNIVERSITY PRESS		电　　话：0335-8387555	
地　　址：河北省秦皇岛市河北大街西段 438 号		邮政编码：066004	
印　　刷：涿州市般润文化传播有限公司		经　　销：全国新华书店	

开　　本：710 mm×1000 mm　1/16		印　张：9	
版　　次：2024 年 9 月第 1 版		印　次：2024 年 9 月第 1 次印刷	
书　　号：ISBN 978-7-5761-0656-5		字　数：165 千字	
定　　价：78.00 元			

编　委　会

主编：

武大勇　　［德］圭窦·库克曼斯特（Giudo Kuchelmeister）　吴军梅

编委：

衡水学院：

张素芳　李思思　石宗琳　李　峰　张　娜

张晨星　耿明雪　郭一丁　王晓超　彭猛威

孙新玲　干海迪　王雅琪

河北衡水湖国家级自然保护区管理委员会：

刘振杰　杜国华　董立文　华　磊　李志芳

王　博　刘蒙蒙　郭亚乐　张余广　侯会军

郑　丽　马晓辉　刘泽泽　孟　静　李　博

王颖琛　李　壮

衡水第一中学：

彭吉栋　武子尧（Matthew Wu）

衡水市第四中学：

徐　倩

目　　录

Directory

第1章 概述

1.1 背景

衡水湖系由古黄河、古漳河、古滹沱河、滏阳河等多条河流数千年摆荡冲刷而成，现湖泊面积为 75 km²，是华北平原单体水域面积最大的内陆淡水湖泊，享有"京津冀最美湿地""京南第一湖"等诸多美誉，国际湿地组织将其喻为"东亚地区的蓝宝石"。

衡水湖国家级自然保护区（Hengshui Lake National Nature Reserve，HLNR），面积 163.65 km²，距北京市、天津市均约 200 km。HLNR 具有涵养水源、净化空气、降解污染、维护生物多样性等重要的生态服务功能，是极具典型性和稀缺性的国家水利风景区，也是东亚 – 澳大利西亚鸻鹬鸟类保护网络成员、国家 4A 级旅游景区、国家生态旅游示范区，被赋予比南方水乡湿地更丰富的内涵和承载。HLNR 生物多样性丰富，是北温带野生动植物聚集地和几百万只候鸟南北迁徙的密集交汇区，有植物 594 种，鱼类 45 种，昆虫 757 种，鸟类 334 种，其中，国家Ⅰ级保护鸟类有丹顶鹤、白鹤、东方白鹳等 21 种，Ⅱ级保护鸟类有大天鹅、小天鹅、鸳鸯、白枕鹤、灰鹤等 63 种。

衡水湖国家级自然保护区为可持续发展教育（education for sustainable development，ESD）活动提供了绝佳的学习场所。ESD 是中德财政合作衡水湖湿地保护与恢复项目（2016 年至 2022 年）的重要组成部分，该项目由德国政府通过德国复兴信贷银行（KfW）资助，其目的是通过在生物多样性保护、生态恢复以及基础设施和能力建设方面的规划、监测和投资，实现衡水湖国家级自然保护区的可持续管理和生态功能保护。

该项目为指导教师制定了一套 ESD 指南，由一系列不同主题的小册子组成。本指南为第一册，主题是"水科学"。因在内容方面侧重自然科学，故而没有按照 ESD 模块的设计标准进行编排。

2019 年，该项目采购了一艘浮桥双体船（因其吃水浅可在浅水中使用），并将其打造为水上教室，面向中小学生开展 ESD 水科学活动。同年，开始对学生和教师进行模块测试。

1.2 目标人群

本指南将在水科学活动的学习单元选择、准备和跟进方面为中小学教师提供帮助。水科学活动的活动场所为水上教室，主要开展对象为四至六年级的小学生和中学生，当然，四年级以下的学生、对自然体验之旅感兴趣的家庭，也可参加在水上教室进行的沉积物或浮游生物的采样活动。此外，水科学活动也适用于在读大学理科生，对于这一目标群体，衡水学院还可以组织更高级的采样活动。

1.3 指南内容

第 1 章：指南简介，包括指南的编写背景、目标人群和内容概要。

第 2 章：水科学活动导论，包括水科学活动的目标、主要特点、研究内容、活动路线和安全管理等。

第 3 章：水科学活动按"学习单元"划分，共分 12 个单元。1 个单元1 个特定主题，根据单元长度和主题，1 个单元可能需要 1 节课（即一个课时）才能完成，也可能需要几节课，甚至更多。

第 4 章：水科学活动数据记录与数据分析。

第 5 章：水科学活动延伸。

本指南还附有数据记录表。

第 2 章 水科学活动导论

2.1 水科学活动的目标和预期效益

2.1.1 水科学活动的目标

1. 为学生提供自然教育，并尽可能让他们接触职业专家，如生态学家、湖泊生物学家、地质学家、化学家、环境和水质工程师。

2. 通过户外的直接体验学习，培养和提高学生对湖泊环境、生物多样性、水质保护的兴趣，以及对可持续发展的兴趣。

3. 证明"户外学习会浪费学生的学习时间，降低他们的考试成绩"的说法是不正确的。

2.1.2 水科学活动的预期效益

预期效益：其他国家的研究表明，学生亲自参与体验大自然的活动，可以拓展他们对世界的理解；通过参与户外活动，学生不仅有直接收获，而且可以获得远超这些直接收获的长远意义。为增强活动效果，水科学课程完全根据各年级学生水平制定，也可以按照个人需求订制。这就是为什么在其他国家，家长愿意为孩子的几小时的水科学活动支付150—300欧元的原因。

1. 实验组（参加 ESD）学生的考试成绩与对照组（未参加 ESD）相同或更高。（信息来源：各校教师记录的学生考试成绩）

2. 与对照组相比，实验组对自然保护的认知水平要高，对建设生态文明的承诺意识也要高。（信息来源：ESD 中心调查的 ESD 活动一年后的数据）

2.2 水科学活动的主要特点

1. 水科学活动是一种完全符合中国综合教育改革的新型教育方式，适用于中小学生。为增强教学效果，教师可以根据各年级学生的水平调整活动难度。

2. 水科学活动的教学重点是让参与者总结在活动中所学的知识，解释所收集的数据。

3. 水科学活动的管理须同时遵守水上安全管理最高国际标准和衡水湖国家级自然保护区的管理规定。水上教室须配备急救箱、救生员和救生绳等，并且保证能够正常使用。

4. 水科学活动配有引导员和助理引导员，引导员主要从能胜任此项活动的衡水学院教师和保护区工作人员中聘用；此外还需要助理引导员，他们负责在船上辅助孩子上课。船上人员及其资质如表 2-1 所示。

表 2-1　船上人员及其资质

职位	资质
引导员	每 10 名学生配备认证引导员和助理引导员各 1 名
船长 / 驾驶员	驾驶员需持有专门的驾驶证，会游泳，并经过安全管理培训
救生员	救生员需持有中国认证的救生员证书，但如果船长或其中一个引导员有此类证书的话，则无须在船上配备专门的救生员

5. 水科学活动引导员职责：

（1）审查从每个站点收集的数据。

（2）比较来自不同站点的数据。

（3）讨论实际值与预期值的一致性和趋势。

（4）明确物理、化学和生物学概念的实际应用。

（5）主持问答环节。

6.水科学活动采集的样品，可由衡水学院和监测站协助完成室内分析。

2.3 水科学活动的研究内容

2.3.1 水质是什么？

水质为评价水体质量状况的基本指标，认识水质以及如何评价水质，对学生来说至关重要。

在航行中，经常有人问"水质是什么？"，在水上教室进行的水科学采样和测量活动可以解答此问题。

水质由水体的物理、化学和生物特性决定，并且与水的特定用途有关。

首先，水质标准与水的用途有关，比如，饮用水标准与划船、娱乐用水标准不同。同一水体，对温水鱼来说是可接受水质，对冷水鱼则不是。其次，水质是水体物理、化学和生物性质的综合体现，不能用单一因素表征。最后，水体的水质也不是一成不变的，会随季节和位置的变化而变化。

2.3.2 水的物理性质

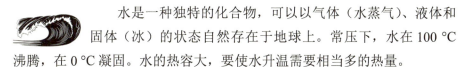
水是一种独特的化合物，可以以气体（水蒸气）、液体和固体（冰）的状态自然存在于地球上。常压下，水在 100 °C 沸腾，在 0 °C 凝固。水的热容大，要使水升温需要相当多的热量。

在水上教室要测量的水的物理性质包括透明度、色度、浊度和温度，可以用塞氏盘、水色计、浊度计（浊度管）和温度计等进行测量。水中的悬浮颗粒会影响水的色度和透明度，当其沉降到湖底会形成沉积物。

2.3.3 水的化学性质

水的化学性质受许多因素影响，例如地质、光合作用、呼吸作用、污染物的负荷、压力、温度等，此外，还会随时间变化而变化。作为溶解物质（溶质）的溶剂，水可溶解离子（带电荷的粒子）也可溶解分子；可溶解气体、固体也可溶解其他液体，被称为"通用溶剂"。当然，也有水不能溶解的物质。

一般情况下，固体在水中的溶解度会随着温度的升高而增大，而气体在水中的溶解度会随着温度的升高而减小。

化学物质的浓度一般用毫克每升（mg/L）和气体饱和百分比表示。在水科学活动中，将会测量水的pH、溶解氧、电导率和碱度等化学性质。

2.3.4 水的生物学特性

湖泊是一个由生产者（水生植物、藻类）、消费者（动物）和分解者（细菌、真菌和一些动物）构成的生态系统，相互之间通过食物网关联并保持平衡，一旦引入外来物种，这种平衡就可能被破坏。

水体的生产力由有效营养物质、光照和温度等因素决定。

水生生物分为以下几类：

（1）浮游生物——随水流漂移的生物：①浮游植物（蓝藻、绿藻、硅藻）；②浮游动物。

（2）游泳动物——能在水层中自由选择其行动途径的水生动物的总称。

（3）底栖生物——生活在湖泊和溪流底内或底表的生物。

（4）分解者——以动植物等生物的遗体、残骸、粪便等为食的那些生物（细菌、真菌和一些动物）。

（5）水生大型植物——除小型藻类以外所有的水生植物类群。

这些生物的空间分布与其所生活的区域（沿岸、开阔水域和底部）有关。

在船上使用浮游生物网采集浮游生物，使用波纳尔抓斗采样器采集底栖生物。

生物量是通过生产者和消费者流动的，衡水湖的生产者包括浮游植物（藻类）和水生植物（大型植物），消费者包括浮游动物、饲料鱼、掠食性鱼类以及以鱼为食的人类和其他动物。衡水湖的食物网由两个不同但相互交叉的部分组合而成（见图 2-1）。

（1）浮游食物网（近岸开阔水域中上层）；

（2）底栖食物网（湖底）。

两个食物网都与浮游植物密切相关。

浮游食物网由藻类等生产者，原生动物、水蚤、桡足类和轮虫等小型无脊椎动物和上层鱼类构成。底栖食物网的能量由藻类、底栖生物、鱼类和光合带掉落的碎屑物（死亡和分解的有机物）提供。

随着捕鱼量的增加和人为造成的环境退化，以浮游生物和底栖生物为食的鱼类，其组成发生了很大变化，鱼类群落的生物完整性已不复存在。

除人类外，衡水湖的鱼类消费者还包括鸟类，如：苍鹭、鱼鹰、潜鸟、鸬鹚和秋沙鸭等。由生产者、消费者与碎屑分解者形成的食物网是一个完整的循环。

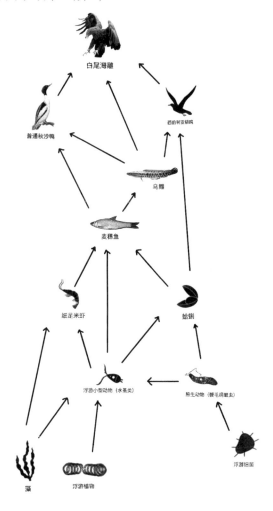

图 2-1　衡水湖的食物网

2.3.5 水质评估

衡水湖面积约 75 km²，与其他湖泊一样，衡水湖每年都要引调黄河水进行补水。近来，南水北调工程贯通，增加了引调长江水补水。

水质是水体物理、化学和生物性质的综合体现，并且与水的用途有关，例如，同一水体对温水鱼来说是"良好"水质，但对人来说却不是。水质评价很复杂，可以通过营养状况或生物生产力对水体进行综合评价。湖泊富营养化或退化的形成需要经过一系列营养状态变化（贫营养—中营养—富营养）。

水体的营养状况或生物生产力可依据营养水平、生物量、溶解氧和透明度等参数来确定。专门为水科学课程设计了水体的营养等级。通过评估各参数的数据来确定采样点水体的营养等级是 O（贫营养）、M（中营养），还是 E（富营养）。

贫营养湖的特征是营养水平低、生物量低、溶解氧高、透明度高；富营养湖的营养水平高、生物量高，底层水的溶解氧低、透明度低，这是因为湖底的沉积物中有大量有机物，它们在分解时会消耗氧气；中营养湖的特征介于贫营养湖和富营养湖之间（见图2-2）。

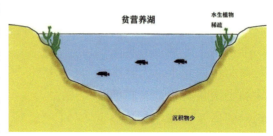

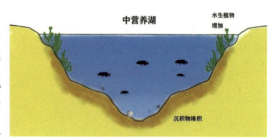

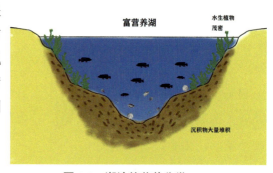

图 2-2　湖泊的营养分类

衡水湖开阔水域、芦苇区和香蒲区的营养状态指数（trophic state index，TSI）在 49.8—62.7 之间，属于富营养水体；小湖的 TSI 较高，在 54.1—74.3 之间，属于富营养水体，个别月份为营养过剩水体（见表 2-2）。

表 2-2　衡水湖 4—11 月的营养状态指数

时间	开阔区	芦苇区	香蒲区	小湖
4 月	49.8	—	—	54.1
5 月	58.1	58.4	57.8	66.9
6 月	55.6	56.9	57.9	61.3
7 月	56.4	60.2	57.4	72.4
8 月	60.9	59.5	61.9	—
9 月	59.4	59.2	60.3	69.5
10 月	62.7	—	—	74.3
11 月	60.5	—	—	69.9

TSI 可以用于评估湖泊或水库的营养状态，划分水体的营养状态级别，它基于多个参数的综合评价，包括叶绿素（Chl）、磷（P）、透明度（SD）等。水体的营养等级与相关参数之间的关系如表 2-3 所示。

表 2-3　水体的营养等级与相关参数之间的关系

营养状态指数（TSI）	叶绿素（Chl）	磷（P）	透明度（SD）	营养等级
$30 < TSI < 40$	$0 \leq Chl < 2.6$	$0 \leq P < 12$	$4 \leq SD < 8$	贫营养
$40 \leq TSI < 50$	$2.6 \leq Chl < 20$	$12 \leq P < 24$	$2 \leq SD < 4$	中营养
$50 \leq TSI < 70$	$20 \leq Chl < 56$	$24 \leq P < 96$	$0.5 \leq SD < 2$	富营养
$70 \leq TSI < 100+$	$56 \leq Chl < 155+$	$96 \leq P < 384+$	$0.25 \leq SD < 0.5$	营养过剩

2.4 水科学活动的采样设备和分析仪器

> 在水科学活动中，体验者要像湖泊学家（研究淡水的人员）一样，使用专业的设备采集水样和湖底物质（沉积物）并进行观察分析。

水上教室备有采集水样和湖底沉积物的专用设备，以及测量水的色度、透明度和温度的专用分析仪器（见图 2-3）。如果水科学活动的体验者在上船之前已经学过湖泊环境检测的基本知识，那他们便了解这些专用设备。但通常人们并不熟悉这些仪器设备，因此在采样之前，会在船上详细介绍这些仪器设备的使用方法。

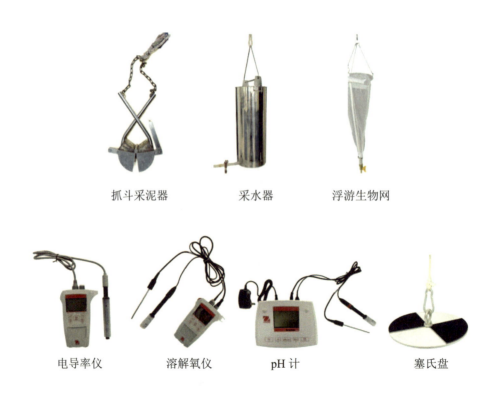

抓斗采泥器   采水器   浮游生物网

电导率仪   溶解氧仪   pH 计   塞氏盘

图 2-3 采样设备和分析仪器

2.5 水科学活动的流程和路线

2.5.1 水科学活动的流程

水科学活动的基本流程包括：到达采样点前讨论采样程序、甲板上的活动以及船舱区仪器的使用；到达采样站；确定采样点位置和深度；采集实验室分析的水样；使用甲板上的设备测量水温，确定水的透明度和色度；采集沉积物样本，在甲板上进行观察；用浮游生物网采集浮游生物，并带回实验室进行显微镜观察（如果船上有特殊显微镜，也可在船上观察）。许多体验活动是同时进行的，如果体验者在出行前已经了解了活动内容，那么他们的体验感将得到最大限度地提升。水科学活动结束时，引导员会组织一个总结会，回顾活动过程并分析采样结果。

2.5.2 水科学活动的路线

图 2-4 是当前最常用的水科学活动路线和采样点，位于衡水湖东北角的实验区内。需要说明的是，在衡水湖国家级自然保护区重新规划分区后，可能会调整路线。

图 2-4 水科学活动常用采样点和路线图

2.5.3 为什么确定采样点的位置很重要？

如果采样位置不确定、不正确或不匹配，水质数据就无法使用，因此确定水样、湖底沉积物的采集位置对后续分析非常重要。

明确采样位置，使得以后可以在同一地点再次采样进行对比，也方便他人找到采样点。

可以根据船的航速、航向和航行时间来推算采样点。但如果考虑风吹、浪大、水急这些因素时，对船只位置判断的准确性就会降低，采用这种方法只能推断一个近似的地点。

水上教室将始终返回到同一个预定地点（见图 2-4）。

2.5.4 如何找到确切的采样点？

导航图用经纬度坐标标记位置。

可以通过全球定位系统（GPS）精准确定采样点坐标。此定位方法将由引导员向体验者展示。

水科学活动结束后，需要在数据表中填写每个采样点的经度和纬度，衡水湖的地理坐标范围为东经 115°28′27″—115°41′54″，北纬 37°31′39″—37°41′16″。

2.6 水科学活动的安全管理

2.6.1 法规和原则

水科学活动将同时遵守中国休闲船航行最高标准和国际标准。至少有一名船员持有救生员证书，并能进行水上有效急救，水科学课程才能进行。

对船长 / 驾驶员和引导员来说，安全是首要问题，这样才能确保水科学活动参与者在船上得到良好照管。在船长和船员当中，至少有一人是中国救生协会（CLSA）认证的救生员，全体船员和教育工作者都应会游泳，通过专门培训。引导员和学生比例如表 2-1 所示。

表 2-4 列出的是与风速相匹配的航行规定。

表 2-4　衡水湖休闲船安全航行规定

风速		规定
3 BFS	12.96—20.37 km/h	高度重视安全，限速航行
4 BFS	20.37—29.63 km/h	水务应急中心红色预警，电瓶船禁止航行
5 BFS	29.63—40.75 km/h	禁止所有船只航行

2.6.2 安全设备

船上的安全设备要符合水科学活动和休闲船的国际标准，如表 2-5 所示。

表 2-5　安全设备

标志	名称（要求）	数量
	救生衣（等级 50S 或更高）	1 人 1 件 *
	锚和锚链 / 锚索（要与船只尺寸 / 重量和水底情况相匹配）	1 套
	水斗 / 水桶 / 消防桶（带绳）	1 件
	舱底泵——手动或电动	1 台 *
	灭火器（装有电动马达、电池、燃气炉或燃料炉的船只）	1 套 **
	船桨或船桨和桨架（除非装有第二种推进装置，6 m 以下的船需配备船桨或船桨和桨架）	1 套
	安全标签（用于容器）	1 种
	声音信号（喇叭 / 口哨 / 铃铛）	1 个

（续表）

标志	名称（要求）	数量
	防水手电筒（可漂浮、可用）	1 个

* 救生衣必须合身，并状态良好可用
** 大船可能需要额外的舱底泵和灭火器

2.6.3 安全乘船程序

船长和引导员都需接受过安全培训。当遇到突发紧急情况时，船长指挥船员和引导员做出响应，大家必须执行船长命令，并充分合作。出于安全考虑，所有人员和背包都要接受检查，不可将背包和其他非必需物品带上船。乘船期间，所有体验者都要穿上能够正常使用的个人漂浮设备（救生衣）。

1. 登船前

水科学引导员做好以下工作。

（1）欢迎体验者，介绍船长和引导员；

（2）做安全讲座：讲解禁区、个人漂浮设备、应急程序及其他安全规则。

2. 登船过程

引导员协助学员穿戴个人漂浮设备。船长（或引导员）介绍船上的各个区域。

（1）船舱，放置存储箱和桌子的地方，存储箱用于存放观察、测量仪器和安全设备，桌子用于书写和放置显微镜等仪器。

（2）禁区、驾驶室和船顶。

3. 起航后

航行中，引导员介绍水科学活动要用到的仪器设备：塞氏盘（测量水的透明度）、福雷尔－乌勒标准色（确定水的色度）、浮游生物网（浮游植物和其他水生生物的采样）、双筒望远镜（观鸟）、显微镜（浮游植物分析）。

2.6.4 给引导员的安全管理提示

1. 一定要提前确定接船的时间和地点，航行起点是衡水湖自然保护区码头。

2. 做好所有天气状况的应急准备。除恶劣的暴风天气（见表 2-4）外，无论天气如何（即使是寒冷的雨天），航行也不能取消。

3. 建议体验者穿着合适的衣服。根据时令建议体验者穿合适的衣服，比如穿长裤还是短裤。湖面上一般会比陆地上冷，而且多风，因此需要带上夹克。此外，因为湿甲板很滑，体验者不能穿凉鞋或拖鞋。而应穿不漏脚趾的橡胶底鞋。对阳光敏感的人还应戴帽子，穿长袖衬衫、长裤子，涂防晒霜。

4. 限制随身携带的物品。船上存储空间不大，航行过程不需要背包，也不能在船上吃东西，须将午餐、背包等物品留在岸上。

5. 确定团队的特殊需求。组织者应知晓所有体验者的特殊需求。如果晕船，在上船前的适当时间服药；如果对蜜蜂、黄蜂等昆虫过敏，或者需要其他特殊药物，可作为随身物品带上船。目前，水科学活动不接受残疾人员。

第 3 章 水科学活动学习单元

　　本手册特意使用"学习单元"一词，1 个单元 1 个特定主题（单元与主题的对应关系见表 3-1）。由于单元的长度和主题不同，有的只需 1 节课（即 1 个课时）即可完成，有的则需几节课，甚至更长时间才能完成。

表 3-1　学习单元主题介绍

编号	标题	章节
学习单元 1	水深测量	3.1
学习单元 2	水样采集	3.2
学习单元 3	水温测量	3.3
学习单元 4	水的透明度测量	3.4
学习单元 5	水的色度测量	3.5
学习单元 6	水的浊度测量	3.6
学习单元 7	水的电导率测量	3.7
学习单元 8	水的 pH 测量	3.8
学习单元 9	水的碱度测量	3.9
学习单元 10	水中的溶解氧测量	3.10
学习单元 11	底泥采集	3.11
学习单元 12	浮游生物采样	3.12

为方便区分，采用通用图标表示检测类别，如下文所示。

物理检测 　　　　　 化学检测 　　　　　 生物检测 　　　　　 数据分析

这些图标也可组合起来使用。

3.1 学习单元 1：水深测量

3.1.1 引导员指南

类型	物理检测	
目标	1. 了解知晓水深重要的原因 2. 学会如何确定水深	
辅助物	带有绳结和锚的特殊绳索、水质数据记录表	

3.1.2 如何确定水深？

知晓水深非常重要，至少有两个原因：

（1）防止船舶搁浅。

（2）可以将科学发现与所采集水样的深度联系起来。

许多水质参数（如温度和溶解氧）会随湖水深度变化，也会随一天中的时间而变化。水的浊度会影响光穿透水的深度，进而影响水生生态系统中的植物和藻类的生产力。所以，湖泊的水深不同，存在的底栖生物也不同。浮游生物和鱼类会随着环境的变化在不同深度的湖水中迁移。

一种简单而古老的测量水深的方法如下：将绳索的一端固定在船侧，另一端系上重物，将重物放入水中，当重物触及底部时，绳索就会松弛，然后将绳索拉回船上，测量从水面到水底的绳索长度，即为水深。

3.1.3 常被问及的问题

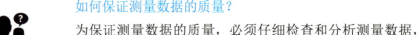

如何保证测量数据的质量？

为保证测量数据的质量，必须仔细检查和分析测量数据，例如：平面控制、高程控制和定位点的准确性，航行障碍物探测的完整性，测深线布局的合理性，深度点的密度以及等高线轮廓图的准确性等。应使用锚点的中值误差来评估锚点的准确性。

3.2 学习单元 2：水样采集

3.2.1 引导员指南

类型	理化检测	
目标	1. 学会如何在船上采集水样 2. 了解什么是 Van Dorn 采样器 3. 采集适量水样	
辅助物	Van Dorn 采样器、水质数据记录表	

3.2.2 如何在船上采集水样？

在船上，将使用两种采样器采集指定深度的水样，一种是常见的 Van Dorn 采样器，另一种是不常见的 Kemmerer 采样器，Van Dorn 采样器适用于各种深度的采样。

3.2.3 什么是 Van Dorn 采样器？

Van Dorn 采样器的主体是一个两端开口的透明塑料圆柱体，该圆柱体连接在水文用金属线（缠绕在绞盘上的钢丝）上，甲板上的工作人员操作绞盘，将 Van Dorn 采样器降到指定深度，即可采集该位置的水样。Van Dorn 采样器上留有安装温度计的接口，装上温度计就可以记录其所在位置的水温。

圆柱体的两端都装有橡胶盖，橡胶盖连在绳子上，可以拉出盖子，将其扭到一侧。将 Van Dorn 采样器降到预定深度停留，直到附载的温度计读数恒定为止。

在将 Van Dorn 采样器放入水中（称为"投放"）之前，需要先确定投放深度，投放深度取决于站点的水深和所需要的样品数量。通常只采集 2 个水样（表面和底部），当需要采集 3 个水样时，可以用 Kemmerer 采样器采集表面水样，用 Van Dorn 采样器采集中部和底部水样。在浅水区，通常只采集 1 个水样。

透过 Van Dorn 采样器的透明壁可以看到水样中的生物。从瓶中取出水样进行分析，并从附带的温度计上读取温度。

3.2.4 常被问及的问题

如何储存水样？

水样的储存方法包括：（1）冷藏或冷冻保存；（2）使用化学试剂保存。例如：在测定氨氮、硝酸盐氮和化学需氧量时，为抑制生物氧化和还原，通常在水样中添加 $HgCl_2$；在测量金属离子时，为防止金属离子的沉淀和吸附，通常在水样中添加 HNO_3 将 pH 调至 1—2。

3.3 学习单元 3：水温测量

3.3.1 引导员指南

类型	物理检测	
学习目标	1. 学会如何在船上测量水温 2. 阐述测量水温的意义，至少说出 3 条 3. 分析衡水湖温度变化的原因和衡水湖全年的温度变化情况 4. 学会如何使用温度计	
辅助物	水质数据记录表、乳胶手套、钟表或手表、足够长（能将温度计放到水中）的细绳、酒精温度计、蒸馏水、500 mL 的烧杯	

3.3.2 如何在船上测量水温？

在船上，通常使用安装在采样器上的温度计来测量水温。在采样器取回后，必须立即读取温度。记录温度时，应以摄氏度（℃）为单位。

3.3.3 测量水温的意义是什么？

利用水温数据可以进行一些有趣的研究，其中最常见的是探究不同水深的温度情况以及温度变化与一年四季的关系。

水温数据还可以显示水温条件是否适合冷水鱼生存。生物的新陈代谢速度、光合作用和分解速度均对温度敏感，鱼类的洄游和产卵行为与水温变化有关。溶解氧也与水温有关，水温越高溶解的氧气越少。

3.3.4 湖水温度一年四季如何变化？

温带地区的大多数内陆湖，在每年的春季和晚秋，湖水温度从湖面到湖底基本一致（见图 3-1）。

在冬季，当湖面被冰覆盖时，冰面下的水温会略高于冰点，并且从湖面到湖底水温逐渐升高，但不会超过 4 ℃（水在 3.98 ℃ 时密度最大，而不是 0 ℃），因此湖水不会从下到上全部冻结。正因为如此，冰才会漂浮在湖面上。冰层的存在可以防止风引起的水上下混合，从而阻止氧气的扩散。

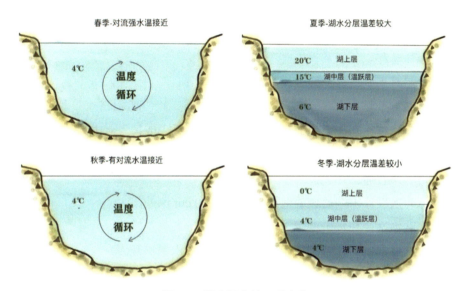

图 3-1　湖水温度的四季变化

在夏季，水温与密度的关系对湖泊状况也起着决定作用。在 3.98 ℃ 以上时，水的密度会随温度升高而降低，较暖、较轻的水会浮在上面。因此，随着夏季太阳热量的增加，湖泊上层的水会变得比下层的更暖、更轻。风的能量不足以将上层水与更冷、更稠密、更深的下层水混合。这种情况导致温暖的上层水与较深较冷的下层水分离，进入夏季分层期。在夏季，湖水存在不同的热温层，此时需为不同的热温层命名以标识其位置。

值得注意的是，夏季湖泊分层会导致风引起的水循环被很大程度上限制在上层水中。由于较低层的水与大气隔离，又几乎接受不到阳光照射，水中溶解的氧气便得不到补充。溶解氧浓度可能会降低到限制水生生物生存的水平。然而，水中溶解的氧气会在夏末得到补充，那时风可以使整个湖泊的水循环起来，湖水温度会变得更加均匀。

衡水湖是一个浅水湖，大多数区域的平均水深在 1.5—3 m，部分区域的水深为 6—8 m。

3.3.5 用酒精温度计测量水温

用品清单

◎水质数据记录表　　◎酒精温度计（带橡皮筋）

◎钟表或手表　　　　◎橡胶手套

操作流程

1. 填写水质数据记录表表头。

2. 戴上手套。

3. 将橡皮筋套在手腕上，以免温度计意外丢失或掉入水中。

4. 检查温度计中的酒精柱，确保液体中没有气泡。如果液线从中间断开，请告知教师。

5. 将温度计末端的玻璃泡插至水样的 10 cm 深处。

6. 让温度计在水中停留 3 分钟。

7. 读取温度，此时温度计的玻璃泡不能从水中移开。

8. 让温度计在水样中再停留 1 分钟。

9. 再次读取温度。如果读数没变，进行步骤 10；如果读数改变，重复步骤 8 直到读数保持不变。

10. 在水质数据记录表上记录温度。

11. 让另外两名学生用新的水样重复测量。

12. 计算 3 次测量的平均值。

13. 所有测量值与平均值的差，都应在 ±1.0 ℃ 之内。如果不是，请重新测量。

3.3.6 常被问及的问题

为什么水温有时比气温低，有时比气温高？

因为水的比热容比空气高，所以与空气相比，水升温或降温需要的时间更长。因此，空气对温度变化的响应比水快得多。

3.4 学习单元 4：水的透明度测量

3.4.1 引导员指南

类型	物理检测	
学习目标	1. 定义水的透明度，并说明其随环境的变化情况 2. 识别塞氏盘，解释塞氏盘读数的意义 3. 正确使用塞氏盘进行透明度测量	
辅助物	水质数据记录表、橡胶手套、塞氏盘透明度野外测量指南、塞氏盘（带绳）、米尺、晾衣夹（可选）	

3.4.2 什么是水的透明度？

水的透明度与光穿透水的深度有关。光进入水体中非常重要，因为太阳是所有生命体的主要能量来源。光是光合作用的必要条件，光合作用是为消费者生产氧气和食物的过程。

通常将透光层（透过的光足以使绿色植物生长的水体上层）的深度视为能见度极限的 2.7 倍（大约 3 倍）。

水的透明度与悬浮颗粒、浮游植物和浮游动物的量有关。

3.4.3 什么是塞氏盘？

塞氏盘是一种非常简单的设备（见图 3-2），直径约为 20 cm，由金属或塑料制成。

塞氏盘是根据对比度来确定能见度极限的。塞氏盘的表面分为四等份，黑白相间。圆盘中心有一个吊环，用于系细绳，以便将塞氏盘从船上或码头上放入水中。塞氏盘的下面附有重物，这样塞氏盘就可以沉入水中。细绳上每隔 0.5 m 有一个标记，这样就可以确定塞氏盘在视野中消失时沉入水中的深度。

图 3-2　塞氏盘

用塞氏盘测量水的透明度的方法属于半定量方法。由于影响测量结果的因素很多，比如，一天中的时间、天空和水面的条件，以及观察者之间的差异等，可能导致在同一位置观察到的塞氏盘从视线中消失的深度不同，这就是为什么在记录塞氏盘读数时要同时记录测量条件信息的原因。

使用塞氏盘测量的标准条件是天空晴朗、阳光直射、波浪或波纹尽可能小。测量必须在船只避光的一侧进行。如果实测条件与上述条件不同，则需要在数据记录中明确说明。值得注意的是，水的能见度约是塞氏深度的两倍。这是因为光照射到塞氏盘后再反射回来，光线经过的长度是塞氏深度的两倍。

3.4.4 塞氏盘读数的意义是什么？

塞氏盘是内陆湖泊监测透明度的标准工具。不同国家和地区的志愿者团体会通过塞氏盘监测湖泊当下的状况，并与往年的数据进行比较。

贫营养或低营养湖泊，塞氏盘的读数通常大于 5 m；富营养化或营养丰富的湖泊，塞氏盘的读数一般小于 2.5 m。

3.4.5 用塞氏盘测量水的透明度

用品清单

◎水质数据记录表　　　　◎米尺
◎橡胶手套　　　　　　　◎晾衣夹（可选）
◎带有绳子的塞氏盘

操作流程

1. 填写水质数据记录表表头。

2. 记录是否有云、云的轨迹类型和云量。

3. 站好，为塞氏盘遮阳，或者用雨伞或硬纸板为测量区遮光。

4. 找一个参照点，它可以是栏杆或码头边缘等，所有测量都以此点作为参照。戴上橡胶手套，因为你可能会碰到被水弄湿的绳子。

5. 将塞氏盘缓慢放入水中，直至其刚好消失。

6. 在细绳上用晾衣夹标记水面位置，如图 3-3（a）所示。如果不能轻易够到水面（例如，站在码头或桥上），请在细绳上标记参照点位置。

7. 将塞氏盘再降 10 cm，然后将其拉起直至其重新出现，如图 3-3（b）所示。

（a）当塞氏盘在水中消失时，用晾衣夹在　　　（b）向上提绳至塞氏盘重新出现，用晾
　　绳上标记水面或参照点位置　　　　　　　　衣夹在绳上标记水面或参照点位置

图 3-3　塞氏盘的使用

8. 在细绳上用晾衣夹标记水面或参照点位置。

9. 现在细绳上应该有两个标记点，在水质数据记录表上记下两个标记点与塞氏盘间的细绳长度，要精确到厘米。如果深度相差超过 10 cm，请重复测量并在水质数据记录表上记下新的测量值。

10. 如果在细绳上标记的是水面位置，请将观察者与水面之间的距离记作"0"。

11. 如果在细绳上标记的是参考点位置，请向下放塞氏盘，直到其到达水面，然后在细绳上标记参考点位置，将标记点与塞氏盘之间的细绳长度记作观察者与水面之间的距离。让其他学生重复步骤5—11两次。

3.5 学习单元 5： 水的色度测量

3.5.1 引导员指南

类型	物理检测	
学习目标	1. 列出影响水的色度的因素 2. 描述什么是福雷尔－乌勒标准色 3. 说明水的色度的重要性 4. 了解如何使用和解释福雷尔－乌勒标准色	
辅助物	福雷尔－乌勒标准色	

3.5.2 什么会影响水的色度？

对水体色度的描述因人而异，许多变量会影响我们对其颜色的感知。其中包括天空条件、一天中的时间点、地面条件、悬浮物质以及观察水的方向。

天空条件包括是否有云。同一水体，在万里无云的晴天呈蓝色（当水中有大量的悬浮物或溶解物以致影响水的颜色时除外），在乌云密布的阴天呈灰色。此外，如果空气中存在大量水蒸气或灰尘，水的颜色也会受到影响。

白天太阳的高度在不断变化，日出日落时分太阳在地平线附近，此时水的颜色会显得很暗。在白天的其他时分，在没有悬浮物质的情况下，水将呈现蓝色或蓝绿色。

同一水面，当它水平如镜时看起来可能是无色的，因为此时水的颜色是水的反射光的体现；当波浪很大且有白色泡沫时，则可能呈现一定颜色，因为此时水的表观颜色还可能会受到穿过波浪的透射光的影响。

除了水本身以外，还有 3 种主要的自然成分会影响自然环境中水的颜色和透明度：

（1）浮游植物，通常为绿色；

（2）溶解的有机物，从黄色到褐色；

（3）其他物质，如细土。从乳白色、灰色到黑色不等。

3.5.3 什么是福雷尔‐乌勒标准色？

因为有许多变量会影响水体的表观颜色，如观察者的感觉，因此有必要建立一种确定颜色的标准方法。有这样一种方法，它通过标准的化学方法制备一系列有颜色的水，这些有颜色的水样被分别密封在不同的玻璃管中，颜色跨度为蓝色—绿色—棕褐，是由铬酸钾、硫酸钴和硫酸铜溶液按一定比例调配而成。这些颜色构成了福雷尔‐乌勒标准色，一种颜色一个编号，1 号是浅蓝色，22 号是棕褐色（见图 3-4）。

在船背阴的一侧将塞氏盘放入水中，下放至水下 1 m 处时，观察者将塞氏圆盘白色部分呈现的水的颜色与玻璃管中的标准颜色作对比，从而确定水的颜色。

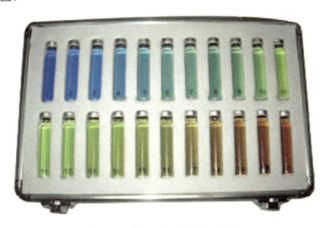

图 3-4　福雷尔‐乌勒标准色

利用上述方法所得到的结果主观性很强，因为对于同一水体，两个观察者测得的结果可能会有一两个数的差别。然而，当每次使用相同的方法来测定水的颜色时，所得数据是可用的。

通过颜色比对测量水质的方法是可靠的。注意：福雷尔－乌勒的标准颜色溶液装在易碎的玻璃瓶中，使用时需要小心，以防其破碎。

3.5.4 测量水的色度的意义是什么？

水体颜色可以反映水的质量和成分。

水体颜色与水中溶解或悬浮的物质有关，也与太阳高度、空气中的水蒸气、云和灰尘等有关（见图 3-5）。

图 3-5　自然界水体的颜色

深蓝色表示水中可能有少量有机物或浮游生物。

绿色表示有浮游植物，并且生物生产力高。

棕色表示水中有矿物质、有机物或大量硅藻。

衡水湖的水呈绿色（见图 3-6），这是由湿地中天然有机物的分解所致。这些有机物包括植物、木材等腐烂产生的单宁酸、木质素和腐殖酸等。

图 3-6 衡水湖水体色度样本

河流中除含有悬浮的黏土矿物质外，还含有有机物，通过浊度测量有助于鉴别使水变色的溶解物质和悬浮物质的种类。

3.5.5 用福雷尔－乌勒标准色测量水的色度

1. 在船的背阴的一侧将塞氏盘下放到水面下 1 m 处。

2. 观察者将塞氏圆盘白色部分呈现的水的颜色与福雷尔－乌勒标准色作对比，记录与福雷尔－乌勒标准色最匹配的颜色编号。

3. 另选两人重复观测，以确定颜色匹配。

4. 解释结果。

3.5.6 常被问及的问题

水的颜色通常分为表观颜色和真实颜色，这是什么意思？

水色是水质的外观指标，分为表观颜色和真实颜色。真实颜色是指去除悬浮颗粒后水的颜色，而带有悬浮颗粒的水的颜色则称为表观颜色。对于洁净或浊度很低的水，其表观颜色与真实颜色相近；对于颜色很深的工业废水和污水，其表观颜色与真实颜色存在很大差异。

3.6 学习单元 6： 水的浊度测量

3.6.1 引导员指南

类型	物理检测
目标	1. 描述什么是浊度 2. 解释如何测量浊度 3. 解释测量浊度的意义 4. 正确使用浊度管
辅助物	水质数据记录表、100 mL 烧杯、橡胶手套、浊度管、采水桶

研究水的透明度的另一种方法是使用浊度计。

3.6.2 什么是浊度？

水的浊度是由各种悬浮物决定的，这些悬浮物可以是有机的（浮游生物、污水），也可以是无机的（淤泥、黏土）。当光射入水中遇到悬浮物时会发生散射，部分光会被悬浮物吸收。光的散射会影响观察者对水的颜色的感知，因此观测到的水的颜色取决于悬浮物的种类和数量。如果将透明容器中的水样放在观察者眼睛和光源之间，就可以看到水的浊度和颜色，这便是使用浊度计时呈现的现象。

3.6.3 浊度是如何测量的？

用浊度计或分光光度计测量水样的浑浊度或不透明度。浊度计由光源、光电管和仪表构成。光路与光电管的朝向成 90°，当光照射水样时，光电管就会检测到样品中悬浮物散射的光，并将其转换成电流，通过仪表显示。依据仪表上指针的位置或数字读数来确定水样的浊度。

可以使用散射比浊法（nephelometric）来测定浊度。浊度单位 NTU 是 "nephelometrie turbidity unit" 首字母的缩写。

在比较不同水样的浊度之前，需要对仪器进行校准。引导员会定期校准浊度计。校准即用仪器测量标准样品，将读数调整到该标准样品的浊度。标准悬浮液通常用福尔马肼配制，这是因为福尔马肼具有稳定的再现性。

3.6.4 测量水的浊度的意义是什么？

浊度与悬浮颗粒对水透明度的影响有关。浊度读数高（清晰度低）表示可能存在侵蚀和沉淀问题。降雨和径流会使河流中的固体悬浮物增加，使河流变浑浊。养分和温度升高会使生物生产效率提高，进而导致硅藻和其他藻类增加，使河流变浑浊，因此浊度计可用于估算浮游生物密度。

衡水湖的浊度约为 15.7—61.2 NTU（见表 3-2），黄河的浊度通常在 20—11 000 NTU 之间。

因为悬浮颗粒会吸收热量，所以浊度升高会导致水温升高。浊度升高会减少射入水中的光，从而导致光合作用速率降低，水的溶解氧含量也随之减少。悬浮颗粒的沉降可能会破坏鱼类的产卵环境和水生无脊椎动物的栖息地，还会堵塞鱼鳃和无脊椎动物的呼吸器官。悬浮颗粒是有害微生物和有毒物质的附着场所。通过絮凝过程可以降低饮用水的浊度，即将明矾或铁、石灰和氯化物的混合物加入水中，使固体沉淀。

表 3-2　4—11 月衡水湖的浊度（NTU）

时间	开阔区	芦苇区	香蒲区	小湖
4 月	26.5	—	—	30.0
5 月	20.6	15.7	22.4	41.4
6 月	33.5	21.8	30.2	27.5
7 月	34.1	28.7	31.4	38.9
8 月	48.8	26.1	31.5	—
9 月	61.2	31.1	40.8	54.1
10 月	51.1	—	—	53.9
11 月	24.6	—	—	34.6

3.6.5 浊度管的使用说明

1. 将水桶中的水倒入浊度管（见图 3-7）中，直至透过水柱看不到管底的黑白图案为止。

2. 一边旋转浊度管，一边自上而下观察管底的图案，确认管底图案的黑白区域是否可辨。

3. 记录这一水深，精确到 1 cm。

4. 登记每个观察者的读数，并计算其平均值。如果给浊度管装满水后，仍能看到管底的图案，则将深度记为"大于（>）管深"即可。

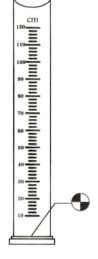

图 3-7　浊度管

3.6.6 用浊度管测量水的浊度

用品清单

◎水质数据记录表　　　　◎ 100 mL 烧杯

◎采水桶　　　　　　　　◎橡胶手套

◎浊度管

操作流程

1. 填写水质数据记录表表头。

2. 记录云的类型、云的轨迹和云的遮蔽情况。

3. 戴上手套。

4. 采集地表水样品。

5. 背对太阳站立，给浊度管遮蔽阳光。

6. 用 100 mL 烧杯将水样缓慢倒入浊度管中。眼睛靠近浊度管口，直视管底，当看不到管底的图案时，停止倒水。

7. 边观察边缓慢旋转浊度管，确保看不到任何图案。

8. 在水质数据记录表上记录浊度管中水的深度，精确到厘米。注意：如果在浊度管装满水后，仍然可以看到管底的圆盘，则将深度记录为 ">120 cm"。

9. 将浊度管中的水倒回样品桶中与剩余的样品混合。

10. 由不同的观察者，用同一水样重复测量两次。

3.6.7 常被问及的问题

水的浊度、颜色和透明度之间有什么区别和联系？

　　浊度是水中悬浮物等对光传播的阻碍程度。由于水中存在不溶物，射入水样中的部分光会被吸收或发生散射。浊度和颜色都属于水的光学性质，但它们并不相同。水的颜色是由水中溶解的物质决定的，而浊度是由水中的不溶物质决定的。因此，一些水样可能色度高但不混浊，反之亦然。 透明度是指水样的清澈程度，纯净的水是透明的。水中悬浮的固体和胶体颗粒越多，透明度越低。通常，地下水的透明度较高。透明度是水质的一个指标，受水体颜色和浊度的综合影响。

3.7 学习单元 7：水的电导率测量

3.7.1 引导员指南

类型	化学检测
目标	1. 描述什么是电导率 2. 描述如何测量电导率 3. 说明电导率的意义是什么 4. 正确使用电导率仪 5. 分析水体电导率变化的原因
辅助物	水质数据记录表、电导率仪说明书、电导率仪、温度计、蒸馏水、两个 100 mL 的烧杯、橡胶手套

3.7.2 什么是电导率？

电导率或传导率是衡量液体导电能力的指标，其数值大小取决于水中离子或带电粒子的数量。依据是否具有导电能力可将液体分为两大类：电解质和非电解质。电流很容易通过电解质含量高的液体，而很难通过弱电解质，例如纯水和一些有机溶剂（酒精或油等）。

3.7.3 如何测量电导率？

电导率仪用于测量水样的导电能力，其通过检测水样中两个电极间的电流来确定电导率。电导率的单位是西门子/米（S/m）或者微西门子/厘米（μS/cm）。

水温越高，电导率越高，水温每升高 1 ℃，电导率增加约 1.9%。一般情况下，测量电导率的标准温度是 25.0 ℃。不同类别的水的电导率范围如表 3-3 所示。

表 3-3　不同类别的水的电导率范围

水的类别	电导率范围（μS/cm）
蒸馏水	0.5 — 2
饮用水	50—1 500
废水	>10 000

3.7.4 测量水的电导率的意义是什么？

电导率测定在水研究中很有用，可以用其估算水中溶解的电解质含量。

电导率值低是贫营养化湖泊的特征。

植物营养物质（肥料）含量较高的富营养化湖泊的电导率值较高。

非常高的电导率值是判断污染地点的良好依据，例如，工业排放物、道路上撒盐和废弃的化粪池都会导致电导率值升高。

电导率发生突变可能表示有污染物通过直接排放或其他方式进入水中。

3.7.5 常被问及的问题

水的电导率和盐度之间有什么关系？

水样中的盐发生电离，以离子状态存在，因而具有导电性。离子浓度越高（即盐度越高），电导率越高，反之亦然。

3.8 学习单元 8：水的 pH 测量

3.8.1 引导员指南

类型	化学检测	
目标	1. 描述什么是 pH，理解酸性、碱性和中性的 pH 差异 2. 说明如何测量 pH 3. 解释测量 pH 的意义 4. 学会使用 pH 计或 pH 试纸 5. 分析衡水湖 pH 变化的原因	
辅助物	用 pH 试纸测量 pH：水质数据记录表、电导率大于 200 μS/cm 的 pH 试纸说明书或电导率小于 200 μS/cm 的 pH 试纸说明书、pH 试纸、50 mL 或 100 mL 烧杯、橡胶手套 用 pH 计测量 pH 值：水质数据记录表、电导率大于 200 μS/cm 的 pH 计说明书或电导率小于 200 μS/cm 的 pH 计说明书、pH 计、蒸馏水、干净的厚纸巾或软纸巾	

3.8.2 pH 是什么？

自然界的水可以呈酸性，也可以呈中性或碱性，其受到多种因素的影响，包括构成蓄水盆地的物质成分，降入水中的雨水的酸度，溪流、河流和暴雨径流汇入水体的状况等。溶液的酸碱度用 pH 表示。pH 的范围在 0—14 之间，当 pH=7 时，水体呈中性；当 pH<7 时，水体呈酸性，由 7 到 0 酸性增强；当 pH>7 时，水体呈碱性，由 7 到 14 碱性增强。由于 pH 是对数值，因此 pH=5 和 pH=6 的水体的酸性差异不是 1，而是 10，即 pH=5 的水体的酸度是 pH=6 的水体的 10 倍。

3.8.3 pH 如何测量？

pH 的测量方法有多种，包括 pH 试纸、pH 笔和 pH 计。

pH 试纸是用指示剂浸泡过的纸条，该指示剂会随酸度变化而改变颜色。将纸条颜色与试纸相应的标准颜色作比对，即可确定 pH。用这种方法测量的 pH 只能精确到 1，但它的成本很低。

pH 笔本质上是一个简单的电极，与 pH 计中的电极类似，都是测量水样中与氢离子浓度相关的电极表面的电势，所测 pH 的精确度范围是 0.1—0.01。

pH 计由测电压装置、浸入水中的玻璃电极、电势恒定的参比电极和温度补偿装置构成。pH 计的读数与温度有关。结果以 pH 单位或毫伏（mV）表示。

pH 试纸和 pH 计：您应该用哪种？

pH 试纸

· 优点：便于儿童使用，不需要校准。

· 缺点：分辨率不如 pH 计，并且没有温度补偿。

pH 计

· 优点：可以测量到 0.1 pH 单位，可以进行温度补偿。

· 缺点：每次使用前必须用缓冲溶液进行校准，比 pH 试纸贵，性能会逐渐变差。

pH 计的使用说明就粘贴在仪器侧面，科学引导员会在使用前对 pH 计进行校准。

3.8.4 测量水的 pH 的意义是什么？

pH 的变化可能与废水排放和污染源有关。然而，在自然条件下，pH 会随二氧化碳水平的变化而改变。二氧化碳在水中的溶解性很强。大气中的二氧化碳和动植物呼吸或分解产生的二氧化碳会进入水中，并与水反应生成碳酸。植物的光合作用会消耗二氧化碳，使地表水的碱性增强。

水质标准中对 pH 的要求通常为 6.0—9.0，当 pH 为 6.7—8.6 时，有助于鱼类种群的均衡发展。当 pH 小于 5.0 或大于 9.0 时，只有极少数鱼类可以耐受（见表3-4）。

衡水湖湖水的 pH 范围通常是 7.7—8.9。

表 3-4　pH 对鱼类和藻类的影响

最小值（pH）	最大值（pH）	影响
4.0	10.1	最具耐受力的鱼类的极限范围
5.0	9.0	大多数鱼类的耐受范围
4.5	9.0	鳟鱼卵和幼虫正常发育
4.6	9.5	鲈鱼耐受的极限范围
4.1	9.5	鳟鱼耐受的极限范围
—	8.7	良好捕鱼水域的上限
5.4	11.4	鱼类应避免出现在超出此范围的水域
6.0	7.2	最适宜鱼卵生存的范围
7.5	8.4	最适宜藻类生长的范围

3.8.5 用 pH 试纸测量水的 pH

用品清单

◎水质数据记录表　　　◎橡胶手套

◎ pH 试纸（电导率大于 200 μS/cm）

◎ 100 mL 烧杯

操作流程

1. 填写水质数据记录表表头。

2. 找到水质数据记录表的"pH"部分，在"pH 试纸"旁的框内打钩。

3. 戴上橡胶手套。

4. 用水样润洗烧杯 3 次。

5. 向烧杯中添加水样至烧杯容积的一半。

6. 按照试纸说明测水样的 pH。

7. 将 pH 数据记录在水质数据记录表上，作为观察者 1 的数据。

8. 使用新的水样和试纸重复步骤 4—6 两次，将测得的 pH 记录在水质数据记录表上，作为观察者 2 和观察者 3 的数据。

9. 计算 3 次观测数据的平均值。

10. 检查测量值与平均值之差，确保其绝对值小于 1 个 pH 单位。如果差值的绝对值不小于 1，需要重复测量。如果重测后仍有测量值不满足此条件，请与教师讨论可能存在的问题。

11. 将用过的 pH 试纸和手套丢入废物容器。用蒸馏水冲洗烧杯。

3.8.6 常被问及的问题

1. 为什么找不到与 pH 试纸匹配的颜色？

一种原因可能是待测水样的电导率较低。当电导率小于 400 μS/cm 时，pH 试纸需要更长的时间完成显色；当电导率小于 300 μS/cm 时，一些 pH 试纸无法正常使用。另一种原因可能是所使用的 pH 试纸因保存时间太久或保存不当而失效。

2. 如果 pH 试纸的颜色似乎介于包装盒上两种指示颜色之间，该怎么办？

选择最接近的 pH。这就是我们为什么要让 3 个学生进行观测的原因，取 3 个读数的平均值可以得到更准确的测量结果。

3.9 学习单元 9：水的碱度测量

3.9.1 引导员指南

类型	化学检测	
目标	1. 描述什么是碱度 2. 说明如何测量碱度 3. 解释测量碱度的意义 4. 使用碱度测定试剂盒 5. 分析水体碱度变化的原因 6. 解释 pH 和碱度之间的区别	
辅助物	碱度测定试剂盒、水质数据记录表、小苏打、碱度说明书、装有蒸馏水的洗瓶、橡胶手套和护目镜	

3.9.2 什么是碱度？

碱度是衡量水中和酸的能力的指标，被称为水的缓冲能力或添加酸时抵抗 pH 降低的能力。水的碱度主要是由于水中的碳酸氢盐（HCO_3^-）、碳酸盐（CO_3^{2-}）和氢氧根离子（OH^-）的存在。它与水中二氧化碳的平衡有关，并且是 pH 的函数。HCO_3^-、CO_3^{2-}、CO_2 间的平衡构成了水的主要缓冲机制。

3.9.3 水的碱度如何测量？

碱度可用酚酞碱度和总碱度表示，这两种碱度均可用硫酸标准溶液滴定至终点时的 pH 来确定。可以使用酚酞和溴甲酚绿－甲基红之类的指示剂来确定滴定终点，也可以用 pH 计来确定。

酚酞碱度滴定终点的 pH 为 8.3，是由全部氢氧根离子和一半的碳酸盐引起的。

总碱度滴定终点的 pH 为 5.1、4.8、4.5 或 3.7，具体 pH 取决于二氧化碳的含量，一般选用 4.5。

碱度的计量单位通常为 mg/L，另一种计量单位是毫克当量每升。

3.9.4 测量水的碱度的意义是什么？

碱度可以用来表示在水中添加酸时，水抵抗 pH 降低的能力。酸的来源通常是雨或雪，不过在某些地区土壤也是酸的重要来源。碱度会随着水溶解含有碳酸钙的岩石而降低。在含有高活性藻类的工业用水中，碳酸盐和氢氧化物都可能对碱度有重要影响。

当湖泊的碱度太低（通常低于约 100 mg/L）时，由强降雨或快速融雪产生的酸会（至少暂时地）消耗掉水中所有的碱，从而致使水的 pH 下降到对两栖动物、鱼类或浮游动物有害的水平。在土壤很少的地区，如山区，湖泊和溪流的碱度通常较低。由于在融雪的第一阶段，污染物往往会从积雪中溶出，因此，在春季通常会有大量酸性污染物进入湖泊，而这时也是水生生物生长的关键时期。

对于高碱度水，可通过使用碳酸氢盐和碳酸盐去除水中的金属，从而降低水的金属毒性。因此，可避免鱼类和其他水生生物吸收这些金属。

3.9.5 用试剂盒法测量水的碱度

用品清单

◎水质数据记录表　　◎蒸馏水
◎碱度测定试剂盒　　◎护目镜
◎橡胶手套

操作流程

1. 填写水质数据记录表表头。

2. 戴上手套和护目镜。

3. 按照碱度测定试剂盒的说明测量水的碱度。

4. 将测量结果记为观察者 1 的数据。

5. 换一份水样，重复测量。

6. 记为观察者 2 和观察者 3 的数据。

7. 计算 3 个测量值的平均值。

8. 每个测量值均应在平均值的可接受范围内。

9. 若某一测量值超出此范围，则舍弃该值并计算另两值的平均值。

10. 如果它们在范围内，则仅报告这两个测量值。

11. 如果有两个以上测量值不在范围内，应从步骤 3 开始重测。

3.9.6 常被问及的问题

T 碱度，M 碱度和 P 碱度有什么区别？

T 碱度是总碱度，是样品中可以被 H^+ 中和的物质总量的量度。M 碱度是甲基橙碱度，是以甲基橙为指示剂时测得的碱度。P 碱度是酚酞碱度，是以酚酞为指示剂时测得的碱度。甲基橙指示剂的变色范围是 3.1—4.4，酚酞指示剂的变色范围是 8—10。由此可以看出，甲基橙碱度大于酚酞碱度，所以在测量总碱度时用甲基橙作指示剂，甲基橙碱度可以被认为是总碱度。以天然水样的测量为例，P 碱度实际上测量的是 CO_3^{2-} 和 OH^- 的含量，而 M 碱度除测量了上述离子之外，还测量了 HCO_3^- 离子的含量。

3.10 学习单元 10：水中的溶解氧测量

3.10.1 引导员指南

类型	化学检测	
目标	1. 描述什么是溶解氧 2. 讲解如何测量水中的溶解氧 3. 描述测量水中的溶解氧的意义是什么 4. 按照溶解氧仪操作规程测量溶解氧含量	
辅助物	溶解氧试剂盒或电极、橡胶手套、护目镜、带盖废液瓶、蒸馏水、温度计、100 mL 量筒、250 mL 带盖聚乙烯瓶、时钟或手表、氧气溶解度表、高程校正表、水质数据记录表、溶解氧试剂盒数据表	

3.10.2 什么是溶解氧？

氧气可以以游离态溶解在水中。

增加：大气中的氧气以及光合作用产生的氧气可能会增加水中的溶解氧浓度。

减少：水中的植物、藻类、动物以及分解有机废物的微生物的呼吸作用会消耗水中溶解的氧气。

水生环境中溶解氧（DO）的分布随水平方向、垂直方向和时间的变化而变化。其分布与大气接触、波浪和水流作用、热现象、废物输入、生物活动以及湖泊或溪流的其他特性有关。

有风时，表层水的溶解氧含量可能会高。溶解氧含量与温度和压力有关，温度低的水比温度高的水溶解的氧气更多。

在白天，光合作用有助于增加水中溶解氧的含量，但水中也存在消耗氧气的生物过程，例如生物的呼吸作用和微生物的有机物分解作用，这些过程消耗的氧气称为生物需氧量（BOD）。

当对氧气的需求量很大，但产生氧气的光合作用未发生时，如日出之前，水中的溶解氧含量可能会很低。沿海地区溶解氧的日波动范围为 4—6 mg/L。在夏天，湖泊深处的溶解氧含量可能会很低。

3.10.3 如何测量溶解氧？

充足的氧气供应是维持水体中生命所必需的，因此氧气含量的测定可以用来评估维持生命的水的质量。目前常用的测定溶解氧含量的化学方法有两种：

温克勒法：向水样中精确添加化学品（试剂），直至颜色发生变化，颜色变化点（或滴定法的其他电子测量点）为滴定终点。

溶解氧仪法：测定的溶解氧的单位是毫克每升（mg/L）。

由于氧气在水中的溶解度与温度、压力和离子浓度有关，因此计算水中氧气的饱和度也很重要。借助溶解氧饱和度的诺模图可以快速估算溶解氧的饱和度值（见图 3-8）。饱和点是指在给定温度下水能溶解的最大氧气量。在给定温度下，当水中的氧气分子含量高于饱和度时，体系处于过饱和状态。晴天光合作用强或水流湍急时，可能会出现过饱和状态。当水样中的溶解氧处于 100％饱和度时为饱和状态，当溶解氧超过 100％饱和度时则为过饱和状态。

如图 3-8 所示，用尺子或深色线，将上面的温度刻度与下面的溶解氧刻度连接起来，标尺或深色线与中间刻度的交点即为溶解氧饱和度。

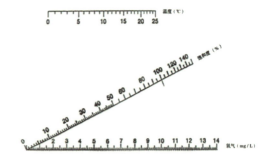

图 3-8　溶解氧饱和度的诺模图

3.10.4 测量溶解氧的意义是什么？

溶解氧水平可以提供在水生环境中发生的生物、生化和无机化学反应信息。大多数水生生物对溶解氧高度依赖，当溶解氧降至约 3 mg/L 以下时，它们将遭受生存压力，甚至会从系统中消失。鲤鱼可在氧气含量为 2 mg/L 的水中生存。

氧气的饱和度低同样也表明水质差。当发生快速生物过程（例如分解或高温）时，溶解氧的饱和度可能会低于 60％。对生物体而言，溶解氧过饱和可能也是问题，因为此时血液中的氧含量会升高，进而导致血液中出现气泡。

溶解氧读数的一般参考：

0—2 mg/L：氧气不足，不能维系生命；

2—4 mg/L：只有少数几种鱼和昆虫可生存；

4—7 mg/L：温水鱼可生存；

7—11 mg/L：非常适合包括冷水鱼在内的大多数溪流鱼生存。

饱和度：

<60％：差，水温过高或细菌耗尽了溶解氧；

60％—79％：大多数水生生物可生存；

80％—120％：非常适宜大多数水生生物生存；

≥ 120％：溶解氧过高，可能危及鱼的生存。

3.10.5 用试剂盒法测量水中溶解氧

用品清单

◎水质数据记录表　　◎蒸馏水

◎橡胶手套　　◎带盖的废液瓶

◎护目镜　　　　　　　◎溶解氧试剂盒

操作流程

1. 填写水质数据记录表表头。

2. 戴上手套和护目镜。

3. 用水样冲洗样品瓶和手 3 次。

4. 给空样品瓶盖上盖子。

5. 将样品瓶浸入水样中。

6. 取下瓶盖，让瓶子装满水，轻摇或轻敲瓶子排除气泡。

7. 在水面下盖上瓶盖。

8. 将样品瓶从水中取出，并倒置检查是否有气泡。如果有气泡，需废弃这个样品，重新采集水样。

9. 按照溶解氧试剂盒的说明测试水样。

10. 将测得的水样的溶解氧含量记录在水质数据记录表中，作为观察者 1 的数据。

11. 让另两名学生分别使用新水样重复测量。

12. 在水质数据记录表中记录观察者 2 和观察者 3 的数据。

13. 计算 3 个测量值的平均值。

14. 3 个测量值与其平均值的差均应在 1 mg/L 以内。如果某一测量值超出此范围，则需计算另两个测量值的平均值，如果这两个测量值与其平均值的差均在 1 mg/L 以内，则记录该平均值。

15. 将所有用过的化学品放入废液瓶中。用蒸馏水清洗溶解氧试剂盒。

3.10.6 常被问及的问题

为什么要在采样点测量水样的溶解氧？

随着外部温度变化和储存时间变化，水样中的溶解氧可能会部分挥发到空气中；空气中的氧气分子也可能会溶解到水中，从而影响测量结果的准确性。因此，测量越快越好。

3.11 学习单元 11： 底泥采集

3.11.1 引导员指南

类型		物理和生物检测
目标	1. 了解湖泊底泥的采样方法 2. 用 Ponar 抓斗式采样器进行采样 3. 初步识别底泥的质地 4. 了解如何研究底泥，以及哪些研究可在船上进行 5. 识别底泥中的几种生物 6. 解释沉积物与营养物之间的联系	
辅助物	Ponar 抓斗式采样器、底栖生物和沉积物数据记录表	

3.11.2 如何采集底泥？

为了粗略分析水体底泥，人们发明了各种装置，其中包括抓斗式采样器、挖泥机、取芯机和钻机。Ponar 抓斗式采样器是船上用于研究底泥组成的主要采样设备。

抓斗式采样器可以用于采集一定数量的未被扰动的底泥。当底泥不太硬也不太软时，它就可以抓取已知区域和深度的样品。正是基于这样的采样方式，因而被命名为抓斗式采样器。

3.11.3 什么是 Ponar 抓斗式采样器？

Ponar 抓斗式采样器由两个相对的半圆形钳口组成，常态下钳口处于打开状态，当采样器下降到湖底，与湖底泥接触时触发开关，驱动一个强力弹簧使钳口闭合，从而将湖底样品抓在斗内。因钳口顶部覆盖有细铜筛网，当采样器被拉回时，抓斗内的物质也不会被冲走（见图 3-9）。

图 3-9　Ponar 抓斗式采样器

工作人员通常会在航行开始时将 Ponar 抓斗式采样器放到甲板上的一个偏僻位置，当船停靠到采样站点需要进行底泥样品采集时，再将其拿出。取样时，采样器处于静止状态，不在水中移动。

采样时，工作人员将 Ponar 抓斗式采样器拿出，并连接到水文测绘线（绞盘绳）上，此时采样器是"待机"状态，即钳口张开，处于待激发状态。然后将采样器悬吊于船侧，并向湖底下放。当钳夹触到湖底时会快速闭合，以抓取湖底样本。只要 Ponar 抓斗式采样器自由悬挂在水文线上，钳口就会保持打开状态；一旦水文线松弛，就会触发开关闭合钳口。当转动绞盘提升 Ponar 抓斗式采样器时，钳口就会闭合，这样便可以从湖底采到样品。当湖水波涛汹涌时，船只会剧烈摇摆，可能导致绞盘绳松弛，从而过早触发开关，使钳口在触到底部前闭合。在这种情况下，须将 Ponar 抓斗式采样器拉回船上，并将开关重置为待激发状态，然后进行第二次采样尝试。

当样品被成功带回到船上后，将 Ponar 抓斗式采样器放入一个长方形不锈钢盒中，该盒底部有一个非常细的网筛。工作人员将 Ponar 抓斗式采样器内的样品倒入该不锈钢盒中，并用软水管冲洗采样器，确保所有样品都被收集到不锈钢盒中。这样，用于分析检测的底泥样品便准备好了。

注：Ponar 采样器由工作人员操作，它很重，所以使用时请注意避让。

3.11.4 如何研究底泥？

研究湖泊底泥有很多种方式。

通过快速目检可以对取回的底泥进行定性分类：沙子、淤泥、黏土、泥土、腐烂的有机物或其混合物。在许多情况下，样品中会存在小动物。用细筛网洗沉积物，可以将生物留在筛子上，将它们挑出放入塑料培养皿中。收集完所有生物后，将培养皿带入主舱，在立体显微镜下进行检查。

可以用分级的细目黄铜筛分离样品，进而研究底部沉积物的成分。沉积物可以按颗粒尺寸进行分类，如表 3-5 所示。

表 3-5 沉积物的分类

沉积物类别	直径 /mm
沙子	2.00—0.05
极粗	2.00—1.00
中等	1.00—0.10
极细	0.10—0.05
淤泥	0.05—0.002
极粗	0.05—0.02
中等	0.02—0.01
极细	0.01—0.002
黏土	< 0.002

沙子具有明显的颗粒形态，很容易观察和感觉到。沙子在潮湿时会形成一定形状，但不会形成带状。黏土在潮湿时具有黏性和可塑性，受到挤压会形成带状。一些沉积物样品的有机物（粪便）浓度较高，这表明这些沉积物的分解速度较慢且氧含量低。沉积物中含有矿物质（例如铁、钙），随着时间的推移，这些矿物质会转化为石灰石、页岩和砂岩。

练习：如何区分沉积物类别？

1. 取一小把沉积物样品。

2. 缓慢加入少量水，与样品充分混合，当样品黏手时停止加水。将土样依次制成不同形状（见表 3-6），直至晾干后不能成形为止。

表 3-6　沉积物的特性

沉积物类别	特性	形状
（1）砂土	松散，只能以单个颗粒堆成金字塔的土壤。	
（2）沙壤土	含有足够的粉粒和黏粒，具有一定黏性，可以团成易碎球的土壤。	
（3）粉壤	可以搓成粗短圆柱的土壤。	
（4）壤土	可以搓成 15 cm 长圆柱的土壤。	
（5）黏壤土	可以搓成"U"形的土壤。	
（6）轻黏土	可以搓成有裂缝圆环的土壤。	
（7）重黏土	可以搓成无裂缝圆环的土壤。	

注：土壤（1）—（4）渗透性好，为砂质、粉质土，土壤（5）—（7）渗透性差，为黏质土。

3.11.5 在底泥中可发现哪些生物？

从衡水湖采集的样品，为观测厌氧降解提供了可能，尤其在 8 月时，生物会耗尽贴近湖底的水中的氧气。衡水湖为底泥研究提供了很多可能。

沉积物中含有大量有机物，有机物腐烂时会释放营养物，当大量有机物腐烂时，就会消耗过多的氧气，从而形成富营养化。从湖岸附近或其他水体底部采集的样品中的物质基本上都是淤泥，其中或多或少有腐烂的有机物，还发现了寡毛纲动物（与蚯蚓相关的分段蠕虫）、蜗牛、水生昆虫和虾。

3.11.6 沉积物与营养物之间的联系是什么？

沉积物是湖泊中营养物的重要来源，沉积物中的有机物腐烂分解时会释放出营养物质。有机物腐烂时会消耗氧气，从而加速富营养化。湖泊的"质量"受许多因素影响，例如湖泊起源和形态、沿岸经济发展、历史污染、相关娱乐项目的数量以及总体水质。

湖边居民上报最多的问题是植物过度生长、藻类大量繁殖和恶臭底泥。这些可能是由水质因素引起的，往往与湖泊管理不善导致的湖泊肥力或生产力提高有关。营养物（氮和磷）含量增加会导致水质和生态系统健康水平下降。营养物可能来自外部（径流、淋滤），也可能来自内部（沉积物）。

香蒲是衡水湖的一种重要挺水植物，它从春季发芽开始就从土壤中吸收磷。对衡水湖香蒲的研究表明，香蒲对磷的吸收在 6 月份最高，然后随着生长而下降，在 10 月份对磷的吸收降到最低（见图 3-10）。10 月的某个时间点，香蒲中的磷会释放回土壤，因此，香蒲在 9 月份的除磷总量会达到峰值（见图 3-11），可在 9 月通过收割香蒲除总磷。

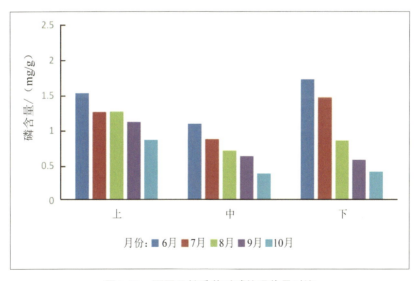

图 3-10　不同月份香蒲对磷的吸收量对比

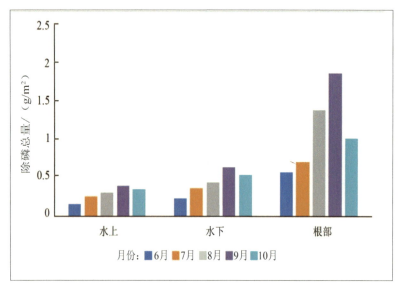

图 3-11　不同收割月份香蒲的除磷总量对比

3.11.7 常被问及的问题

研究湖泊沉积物的意义是什么？

比较不同年代的沉积相有助于了解湖区的古地理。研究沉积物的矿物组成和分布特征，确定沉积物来源，可以为寻找湖相沉积矿床提供依据。湖泊沉积物的厚度和性质可用于确定湖盆的年代，推断沉积物形成过程的水文和气候条件。沉积物中积累了大量有机物和各种稀有元素，为形成各种湖泊沉积提供了物质来源。

3.12 学习单元 12：浮游生物采样

3.12.1 引导员指南

类型	生物检测
目标	1. 描述什么是浮游生物并列出其类型 2. 用浮游生物网捕获浮游植物和浮游动物，并用野外显微镜进行识别 3. 识别一些衡水湖的典型浮游生物
辅助物	浮游生物网或采样器、显微镜、浮游生物数据记录表和典型浮游生物图

3.12.2 什么是浮游生物？

浮游生物分布在整个湖泊中，在各种深度都能找到它，包括植物（浮游植物）和动物（浮游动物）两种形式存在。浮游生物的分布模式与季节和一天中的时间相关。

浮游生物有三种基本体形，如表 3-7 所示。

表 3-7   根据浮游生物的体形大小分类

类型	尺寸	描述
微型浮游生物	5—60 μm	由于体形小，微型浮游生物会穿过标准采样网的孔，专用的细网可用来捕获较大的微型浮游生物
小型浮游生物	2 mm	大多数浮游生物属于小型浮游生物或网采浮游生物
大型浮游生物	最长可达数米	肉眼可见

通过细网过滤水来收集生物。在船上用浮游生物网收集小型浮游生物。

3.12.3 浮游生物是如何取样的？

浮游生物网（见图 3-12）或采样器是收集浮游植物和浮游动物样品的设备。为了对不同样本进行定量比较，有些网装有流量计，用于确定通过收集网的水量。

图 3-12　浮游生物网

浮游生物网或采样器可以获取各种深度的浮游生物样品，从而可以研究浮游生物的分布模式。可通过采样水柱的深度进行定量测定。浮游生物网可以在某一深度水平拖网对浮游生物进行取样（水平拖），也可以将网放到水中向上拖网进行取样（垂直拖），还可以斜拖，即将网降到预定深度，在船向前移动的条件下以恒定速率向上拖网取样。

3.12.4 浮游生物样品中通常会有什么？

衡水湖食物链的底层是浮游生物。其中，浮游植物是生产者，通常有绿藻、蓝藻（以前称为蓝绿藻）和硅藻。蓝藻，如微囊藻，适宜在温水和高营养条件下生存，一些蓝藻可以固定氮并产生毒素。甲壳类动物如水蚤、桡足类等是样品中的消费者或浮游动物代表。

显微镜操作方法：

1. 调节亮度，由暗调亮，调节反光镜的角度。

2. 将临时载玻片固定在载物台的适当位置上。

3. 将低倍物镜对准通光孔，用粗准焦螺旋自上而下调节镜筒，从侧面观察，防止物镜镜头接触到载玻片而损坏镜片或压碎载玻片。

4. 左眼通过目镜观察视野的变化，同时调节粗准焦螺旋，使镜筒缓慢上移，直至视野清晰为止（见图 3-13）。

5. 如果视野中没有被观察对象，可以移动载玻片，原则为欲上反下，欲左反右。

6. 如果不够清晰，可以用细准焦螺旋进一步调节。

7. 如果需要在高倍物镜下观察，可以转动转换器调换物镜。如果视野较暗，则可以通过步骤 1 进行调整；如果不够清楚，可以通过步骤 6 进行调整，但不能通过步骤 4 进行调整。

8. 使用完毕后，调节转换器使空镜头孔对着通光孔，竖起反光镜，将镜筒调至最低后装入镜箱。

图 3-13　显微镜的使用

3.12.5 常被问及的问题

什么是湖泊富营养化？湖泊富营养化的危害是什么？

湖泊富营养化是指因湖水含有过量的氮、磷和其他营养物质，导致水中藻类和其他浮游生物过度繁殖、溶解氧含量下降、水质恶化、鱼类和其他生物大量死亡的现象。

富营养化的危害包括散发出腥臭味、水体的透明度和溶解氧含量降低、向水体释放有毒物质、含有有害的硝酸盐和亚硝酸盐等。

第4章 水科学活动数据记录与数据分析

4.1 数据记录

为水上教室的水科学体验准备了一个数据表，表中列出了所有待测参数的信息。课程结束后，在 Excel 电子表格（见附录）中录入以下数据：每个采样点的采样日期、位置、纬度、经度和深度，顶部和底部的浊度、电导率、温度、pH 和溶解氧，以及塞氏盘读数、福雷尔 - 乌勒色标编号、底栖生物清单、浮游生物相对密度和沉积物类型。进行碱度测量时，也将其录入。

4.2 数据分析

1. 利用在船上收集的各个采样点的温度，绘制成温度曲线。
2. 分析温度曲线，指出异同，并试着解释这些异同点。
3. 制一张表，展示每个采样点的底栖生物的类型和相对丰度。
4. 制一张表，展示每个采样点的浮游生物的类型和相对丰度。

4.3 水生生态系统分析

1. 在所收集的全部数据中，列出两个水生生态系统之间相似性最大的数据和差异性最大的数据，分析这些数据，并用自己的语言表述为什么你认为这些水生生态系统是不同的。

2. 衡水湖采样点的营养状态（富营养、中营养、贫营养）是什么？

4.4 食物链分析

1. 绘制每个采样点的浮游食物链，比较它们的异同点。

2. 绘制每个采样点的底栖食物链，比较它们的异同点。

第 5 章 水科学活动延伸

体验者需要在水科学之旅结束时总结经验。如果条件允许，还应强调诸如气候变化之类的可持续性和挑战性问题。后续室内工作也可以在 ESD 教育中心或学校进行。最好在体验结束时作简要总结，并在学校进行长期跟进。

5.1 数据维护

数据维护是水科学活动的正常延伸。现已有一名统计专业学生分析了多个地点的一年的水质数据，并研究了其发展趋势，本项目可以使用这些数据进行假设检验，了解水质参数是如何随位置和季节变化的，生成各种统计量度、图表和图形。

国家资源计划局（NRPB）的监控部门维护学生数据库，并计划在生物多样性数据库网站的主页上提供在线下载数据。

5.2 留给体验者的思考题

1.华北最大的两个湖泊是哪两个？其中哪一个面积较小？

2.你家的水从哪里来？废水流向何处？

3.当发电厂烧煤时，如何防止煤灰从烟囱逸出？什么物质会从烟囱中逸出？为什么电厂从湖中取水？当水回流到河流或湖泊时，必须采取哪些预防措施？发电厂利用回收石灰做什么？

4.港口、房屋、休闲区、垂钓等沿岸发展对水生生态系统有何影响？

衡水湖国家级自然保护区为促进旅游业做了哪些工作？游客对该地区有何影响？

5. 描述衡水湖的底部沉积物，这些沉积物为什么是这样的？这些沉积物中有什么生物？它们是如何在如此低的氧气水平下生存的？

6. 在几千年前，黄河蜿蜒穿过衡水湖国家级自然保护区，今天你是怎么能看出这一点的，为什么以前的河道遗址是制砖的绝佳场所？

7. 什么是生物多样性？衡水湖中生活着哪些生物？

8. 你会在衡水湖找到什么种类的鱼。为什么？溶解氧是如何对鱼类产生影响的？

9. 纯净水的浊度是多少（以 NTU 为单位）？一年中衡水湖的水什么时候浊度最低？为什么？

10. 哪些因素会导致电导率读数升高？我们期望的衡水湖电导率范围应是多少？

附　录

水质数据记录表

表1:

姓名：_____　　学校：_____

引导员：_____　　日期：_____　　年级：_____

编号水深/cm	水温/°C				透明度（绳长/m）				水色				浊度（水深/cm）				电导率/（μS·cm⁻¹）	pH				碱度/（mg·L⁻¹）				溶解氧/（mg·L⁻¹）			
	1	2	3	平均	1	2	3	平均	1	2	3	平均	1	2	3	平均		1	2	3	平均	1	2	3	平均	1	2	3	平均
1																													
2																													
3																													
4																													
5																													
6																													
7																													
8																													
9																													
10																													

（pH 试纸□，pH 计□）

纬度：_____ N　　经度：_____ E　　时间：_____ 上午/下午

空气温度 °C　　风向：_____　　风速：_____ km/h

表 2:

底栖生物和沉积物数据记录表

纬度:＿＿＿＿＿N 经度:＿＿＿＿＿E 水深:＿＿＿＿＿m 水温:＿＿＿＿＿℃

沉积物类别:＿＿＿＿＿ 放绳长度:＿＿＿＿＿m 采样次数:＿＿＿＿＿

采样人:＿＿＿＿ 记录人:＿＿＿＿ 采样时间:＿＿年＿＿月＿＿日＿＿时＿＿分

编号:＿＿＿＿＿

优 势 种 类 记 录				
序号	种名	总个数 /ind	取回个数 /ind	备注
1				
2				
3				
4				
5				
6				
7				
8				
9				
10				
11				
12				
13				
14				
15				
16				
17				
18				
19				
20				
21				
22				
23				
24				
25				
26				
27				
28				

表3：

浮游生物数据记录表

纬度：_____N 经度：_____E 水深：_____m 水温：_____℃

采样人：_____ 记录人：_____ 采样时间：____年____月____日____时____分

编号：_____

采样项目		瓶号	绳长/m	倾角/（°）		流量计		备注
				开始	终了	编号	转数/r	
拖网	网							
	网							
	网							
	网							
	网							
	网							
	网							
采水							采水量/cm³	
	层							
	层							
	层							
	层							
	层							
	层							

Chapter 1 Introduction

1.1 Background

Hengshui Lake has been eroded by several rivers including the ancient Yellow River, the ancient Zhang River, the ancient Hutuo River, and the Fuyang River for thousands of years. The lake now covers an area of 75 square kilometers. It is the largest single inland freshwater lake in the North China Plain by water area. It enjoys many good reputations such as "Jing-Jin-Ji's most beautiful wetland"and "Jingnan first lake". The International Wetlands Organization calls it "Sapphire in East Asia".

Hengshui Lake National Nature Reserve (HLNR) covers an area of 163.65 square kilometers and is located over 200 kilometers away from both Beijing and Tianjin. HLNR has important ecological service functions such as conserving water sources, purifying air, degrading pollution, and maintaining biodiversity. It is a highly typical and scarce national water conservancy scenic spot, as well as a member of the East Asia-Australasia Snipe Bird Protection Network, a national AAAA level tourist attraction, and a national ecological tourism demonstration zone. It is endowed with a heavier connotation and carrying capacity than southern water town wetlands. HLNR has rich biodiversity and is a gathering place for wild animals and plants in the northern temperate zone, as well as a dense intersection area for millions of migratory birds moving north and south. There are 594 species of plants, 45 species of fish, 757 species of insects, and

334 species of birds. Among them, there are 21 species of national first-class protected birds such as red crowned crane, white crane, and Oriental white stork, and 63 species of second-class protected birds such as whooper swan, cygnet, mandarin duck, white naped crane, and gray crane.

The Sino-German Hengshui Lake Conservation and Management Project (2016 - 2022) cofinanced by the German Government (BMZ) through KfW aims to sustainably manage the ecosystems of the Hengshui Lake National Nature Reserve and conserve its ecosystem services through planning, monitoring, investment in biodiversity conservation, eco restoration and infrastructure and capacity building. Education for Sustainable Development (ESD) is an important project component.

The Project has elaborated an ESD guide for teachers consisting of a series of booklets. This booklet number 1 is entitled "Water Science". It is not following the standard of the design of ESD modules, since the subject matter is more science oriented.

In 2019, a pontoon catamaran was procured, to make a floating classroom to teach water science, targeting primary and secondary school students of the ESD program. Testing of the modules with pupils and teachers has started in the same year.

1.2 Target group

This guide will assist primary and secondary school teachers to actively participate in selection of learning units, preparation and follow up of the outdoor educational event. The learning units are appropriate for fourth through twelfth grade classes. However, sampling of bottom material or plankton can be also done by younger pupils, or interested families taken part in a nature experience trip on the project catamaran. The designed water science activities are also suitable for students enrolling in science in a university like the HSU. The

Hengshui University can also organize more advanced sampling for this target group.

1.3 The booklet comprises

Chapter 1: Introduction, including the background, target group and content summary.

Chapter 2: Introduction to the water science activities, including the objectives, main features, research contents, activity routes and safety management of the water science activities.

Chapter 3: Learning units of the water science activities. This part is divided into12 units based on the "learning units". Each unit has a specific theme, and depending on the length and theme of the unit, it may take one class (i.e. one lesson) or even more classes to complete.

Chapter 4: Data recording and analysing of the water science activities.

Chapter 5: Extension of the water science activities.

The booklet also contains data record sheets.

Chapter 2 Introduction to the water science activities

2.1 Objective and expected benefits of the water science activities

2.1.1 The objectives of the water science activities

1. To provide students with real-world education and exposure to possible careers, such as ecologist, lakes biologist, geologist, chemist, environmental and water quality engineer.

2. To build and enhance the interest in learning about and caring for the lake environment, biodiversity and water quality conservation and sustainable future through direct experience practical outdoors learning.

3. Proving that the statement' outdoor learning wastes students' study time and lowers their exam scores' is wrong.

2.1.2 The expected benefits of the water science activities

Expected benefits: The researches from other countries suggest that students who participate in hands-on experiences in nature will broaden their scope of understanding of the world and their place in it. The benefits of students' outdoor activities far exceed the results of the activities. The onboard lessons can be fully aligned with curricula for each grade level and to customize individual trips to reinforce teachers' classroom curricula.This is one reason why in other countries parents are willing to pay Euro 150 to 300 per a couple of hours for the floating classroom scheme event per pupil.

1. The exam scores of the experimental group (participating in the ESD scheme) of the pupils are the same or even higher than the control group (not participating in the ESD scheme). (Source of information: record of exam scores by respective school teachers)

2. The knowledge towards nature conservation and the commitment to build an ecological civilization in the experimental group are higher than control group. (Source of Information: ESD Center survey after the ESD event; and one year later)

2.2 Key features of the water science activities

1. The water science activities' innovative style of education is fully in line with the Chinese comprehensive education reform. Working with teacher the cruise can be fully aligned with the curricula for the grade level to reinforce teacher's classroom curricular. Aquatic science course is targeting primary and secondary school students.

2. An important element of the water science activities is the summary of what participants have learned in activities and the interpretation of the data they have collected.

3. The water science activities complies with the highest international standard for water safety management also complies with the protected area regulations in HLNR. All security-requirements and needs for first-aid, rescue-swimmer and rescue-chain to run such programs must be in place.

4. Instructors will be mainly contracted from the Hengshui University and protected area staff qualified for such a task. In addition, assist to the instructor will be onboard for supervision of kids for school classes. Onboard persons and their qualification see Table 2-1.

Table 2-1 Onboard persons and their qualification

Position	Qualification
Science instructors	One certified instructor and assistant instructor for 10 students
Captain / driver	One driver with special driving license and swimming capability and training in safe management
Life guard	With a Chinese certified lifeguard certificate, there is no need for a professional life guard on board if the capitan or an instruction has such certification

5. The science instructors of the water science activities will facilitate:

(1) Reviewing the datum gathered from each station;

(2) Comparing the values from the different stations;

(3) Discussing the consistency and trend between actual and expected values;

(4) Identifying the practical applications of physical, chemical, and biological concepts;

(5) Holding the question and answer session.

6. The samples collected by the water science activities can be assisted by Hengshui University and the monitoring station to complete the laboratory analysis.

2.3 The research content of the water science activities

2.3.1 What is water quality?

Water quality is the basic indicator for evaluating the quality status of water bodies. It is very important for students to know water quality and how to evaluate it.

A frequently asked question on cruises is, "What is the quality of the water?" Many of the sampling procedures performed on the vessel help to answer this question.

Water quality is defined in terms of physical, chemical, and biological parameters with respect to a certain use.

Firstly, water quality standards are relater to the use of water. For instance, acceptable water quality for warm water fishes would not be optimal for cold water fishes, and standards for drinking water differ from those for boating and recreation. Secondly, water quality is defined in terms of physical, chemical, and biological parameters. No single factor alone indicates good water quality, and water quality in a body of water can vary with the season and location.

2.3.2 Physical properties of water

Water is a unique chemical compound that naturally exists on earth in the gaseous (water vapor), liquid, and solid (ice) states. Water boils at 100 °C and freezes at 0 °C. A relatively large amount of heat is needed to raise water temperature.

Physical properties of water that are measured on the vessels include water transparency or clarity, color, turbidity, and temperature. Suspended particles in water will influence water color and clarity. Particles can settle to the bottom and contribute to a build-up of sediment. Instruments and equipment used to quantify these properties include the Secchi disk, Forel-Ule Color scale, turbidity meters (turbidity tubes), and thermometers.

2.3.3 Chemical properties of water

 Chemical properties of water is influenced by many factors such as the geology of a region, photosynthesis and respiration, pollutant load, pressure, temperature, and can change with the time of day. Behaving as a solvent in which a substance (solute) can be dissolved, water has been called the "universal solvent". The resulting solution may contain individual ions (particles with charges) or molecules. Gases, solids, and other liquids can be dissolved in water but some of them do not be dissolved, they are insoluble.

The solubility of a solid in water generally increases with temperature, while the solubility of a gas decreases with temperature.

Concentrations of a chemical in water are generally expressed in terms of milligrams per liter (mg/L) and percent saturation for gases. Chemical properties of water investigated by the water science activities include pH, dissolved oxygen, conductivity, alkalinity, and nutrients.

2.3.4 Biological properties of water

 Lakes are ecosystems composed producers (aquatic plants and algae), consumers (animals), and decomposers (bacteria, fungi, and some animals) which are interrelated through food webs. Introduction of exotic or nonindigenous species can upset the balance of existing food webs.

The productivity of a body of water is dependent upon variables such as the available nutrients, light, and temperature.

Aquatic organisms are divided into groups that include:

1. Plankton – organisms that drift with the currents: ① phytoplankton (blue algae, green algae, diatoms); ② zooplankton-animals;

2. Nekton – larger size organisms that can swim freely;

3. Benthos – organisms that live in or on the bottom of lakes and streams;

4. Decomposers –Those organisms (bacteria, fungi, and some animals) that feed on the remains, debris, feces, etc., of animals, plants, and other living things;

5. Large aquatic plants – all groups of aquatic plants except for small algae.

These organisms have a spatial distribution that is defined by regions adjacent to the shore (littoral), open waters (limnetic or pelagic), and the bottom (benthos). On the vessels, plankton nets are used to sample for plankton and a Ponar grab sampler is used to collect benthic organisms.

The general flow of biomass in the Hengshui Lake is through trophic levels that include the producers (phytoplankton/algaeand aquatic plants/macrophyte) and consumers (zooplankton, forage fishes, predator fishes, fish-eating humans and other animals). The Hengshui Lake food web is composed of two distinct but overlapping parts (Figure 2-1):

1. The pelagic food web associated with offshore open water (pelagic).

2. The bottom (benthic) food web.

Both webs are dependent upon the phytoplankton in the surface waters.

Members of the pelagic food web include producers like algae, small invertebrates such as protozoon, cladocerans, copepods, rotifera, and pelagic fish. The benthic food web is fueled by the algae, benthos, fish, and detritus (dead and decomposing organic matter) that fall from the upper part of the water column (photic zone).

The biological integrity of the fish community that is dependent on pelagic and benthic species is no longer present. Increasing levels of fishing pressure and human-induced environmental degradation have greatly altered the composition of fish species.

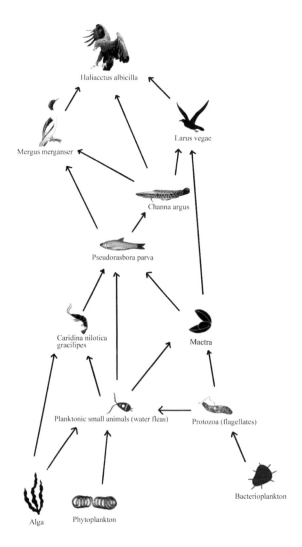

Besides human beings, consumers of Hengshui Lake's fish include birds such as herons, osprey, mergansers. The web comes full circle with the producers, the consumers, and the detritus decomposers completing the cycle.

2.3.5 Evaluation of water quality

The surface area of Hengshui Lake is about 75 km². The Hengshui Lake is a shallow artificial lake. It is replenished annually with water from the Yellow River like other lakes. Recently, the South-to-North Water Diversion Project has been completed, increasing

Figure 2-1 The food web of Hengshui Lake

water diversion from the Yangtze River to replenish water.

Determining water quality is complex; many parameters go into its determination. Evaluation of water quality depends on the designated use of the water. For example, "good" water quality for warm water fishes may not be "good" water quality for human consumption. Overall water quality can be evaluated by considering the trophic status or biological productivity. Eutrophication, or aging of lakes need to go through various trophic states (oligotrophic → mesotrophic→eutrophic).

Nutrient levels, organic matter content, dissolved oxygen levels, and water transparency provide clues to the trophic state or biological productivity of a water body. A trophic scale has been specially designed for use on the water science activities. By evaluating data from various parameters, sampling locations are rated as O (oligotrophic), M (mesotrophic), or E (eutrophic).

Oligotrophic lakes are characterized by low nutrient levels, low biomass, high oxygen concentrations, and high trans-parency.Eutrophic lakes are highly productive with high nutrient levels, high biomass, low oxygen concentration in the bottom waters, and low transparency. The large volume of organic matter accumulated in bottom sediments depletes oxygen as it decomposes. Meso-trophic lakes are between the other two trophic states in their characteristics (Figure 2-2).

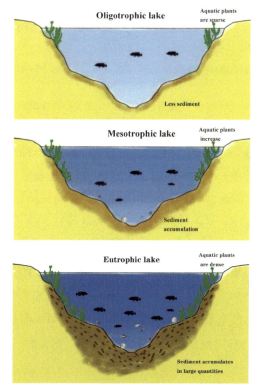

Figure 2-2 Trophic classification of lakes

Open surface waters, reed zones and typha zones mainly belong to eutropher, the trophic state index (TSI) in which ranges from 49.8 to 62.7. The small lake belongs to eutropher, the TSI in which is higher, ranging from 54.1 to 74.3, even hypereutropher in several months (Table 2-2).

Table 2-2 The trophic state index (TSI) in Hengshui Lake from April to November

Date	Open zone	Reed zone	Typha zone	Small lake
April	49.8	—	—	54.1
May	58.1	58.4	57.8	66.9
June	55.6	56.9	57.9	61.3
July	56.4	60.2	57.4	72.4
August	60.9	59.5	61.9	—
September	59.4	59.2	60.3	69.5
October	62.7	—	—	74.3
November	60.5	—	—	69.9

The TSI is a classification system designed to rate water bodies based on the amount of biological productivity they sustain. The TSI of a water body is rated on a scale from zero to one hundred. The Table below demonstrates how the index values translate into trophic classes.

Table 2-3 The relationship between the trophic classification and the relevant parametes

TSI	Chl	P	SD	Trophic Classification
$30 < TSI < 40$	$0 \leqslant Chl < 2.6$	$0 \leqslant P < 12$	$4 \leqslant SD < 8$	Oligotrophic
$40 \leqslant TSI < 50$	$2.6 \leqslant Chl < 20$	$12 \leqslant P < 24$	$2 \leqslant SD < 4$	Mesotrophic
$50 \leqslant TSI < 70$	$20 \leqslant Chl < 56$	$24 \leqslant P < 96$	$0.5 \leqslant SD < 2$	Eutrophic
$70 \leqslant TSI < 100+$	$56 \leqslant Chl < 155+$	$96 \leqslant P < 384+$	$0.25 \leqslant SD < 0.5$	Hypereutrophic

2.4 Sampling equipment and analysing instruments of the water science activities

In the water science activities, participants act as limnologists, using equipment to sample and analyze water and bottom material (sediment).

Participants aboard the Hengshui Lake cruises will receive a basic background in the use of limnological sampling equipment, and analytical instruments (Figure 2-3) and procedures. It is also assumed that before embarking they are already taught the basics about environmental monitoring of a lake. Since people are often not familiar with the specialized equipment and instruments carried on the vessel, detailed instructions on their use are given on-board before samples are taken. A variety of equipment for obtaining water and bottom sediment samples, as well as biological samples are carried on-board the vessels. Specialized equipment for water color, clarity, and temperature measurements is found on the vessels.

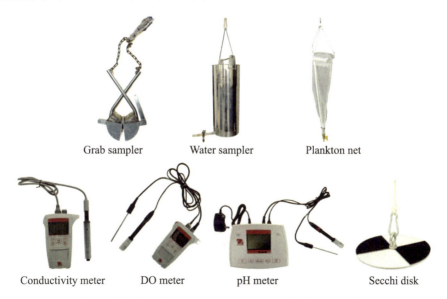

Grab sampler   Water sampler   Plankton net

Conductivity meter   DO meter   pH meter   Secchi disk

Figure 2-3 Sampling equipment and analytical instruments

2.5 The process and route of the water science activities

2.5.1 The process of the water science activities

Discussions include navigation, sampling procedures, side deck activities, and use of instrumentation in the cabin area. The basic flow of the trip is to arrive at a sampling station, determine the location and depth, take water samples for laboratory analysis, measure water temperature, determine water clarity and color using side deck equipment, take sediment samples for deck viewing, and perform a plankton tow followed by microscopic viewing in the laboratory (or if special microscopes are available on board). Many activities are happening at once and participants will maximize their experience if they have been briefed on the activities prior to the trip. There is a wrap-up session at the end of the trip where the science instructors review and interpret the sampling results.

2.5.2 The route of the water science activities

Figure 2-4 shows the most commonly used stations and routes currently located in the experimental zone at the northeast corner of Hengshui Lake. In case in the rezoning of the HLNR, the route may be adjusted.

Figure 2-4　Map of common sampling points and route for the water science activities

2.5.3 Why is the exact location of the sampling site important?

Water quality data are not useful if the sampling location is unknown, incorrect, or mismatched. It is important to know the location of the site where samples of water or bottom sediment samples are taken for analysis.

The location provides information that makes it possible for other samples to be taken to make comparisons at the same place at a later time, and for others to find the site.

It is possible to return to the general area of a sampling station in a lake by using dead reckoning. This is done by keeping track of how fast the boat moves, the time it is moving a given speed, and the direction or directions traveled. If the wind is blowing, the waves are large, a current is present or any combination of these factors, the accuracy of knowing where the vessel is located is reduced. Returning to a given point on the lake by using this method can be only approximate.

The floating classroom cruise will always return to the predetermined sites (Figure 2-4).

2.5.4 How to locate the exact sampling location?

Navigational charts use latitude and longitude coordinates to mark positions.

Location coordinates are pinpointed through use of the Global Positioning System (GPS).The instructor will show how to use GPS for exact location.

The Vessel data sheets have blank spaces for latitude and longitude for each sampling station. Typical reading for the Hengshui Lake is 115°28′27″ - 115°41′54″E longitude and 37°31′39″ - 37°41′16″N latitude.

2.6 Safety management of the water science activities

2.6.1 Regulations and principles

The highest Chinese nautical standards for recreational boating and international standards for floating classes will be applied. Without at least one crew with a national life safety certificate and a valid first aid floating classes should not be permitted to make a cruise.

Safety must be the Captain/driver of the vessels and instructors' number one concern. It will be ensured that the participants are in good hands when onboard. On board are at least the captain or one crew staff is China Life Saving Association (CLSA) certified lifeguard, Crew staff and educators should have swimming certificate. Ratio of instruc-tors to students see Table 2-4.

Table 2-4 indicates the sailing restriction according to wind speed.

Table 2-4　Regulations on safe navigation of leisure boats in Hengshui Lake

	Wind speed	Remark
3 BFS	12.96-20.37 km/h	Pay high attention to safety, limit speed of sailing
4 BFS	20.37-29.63 km/h	The water emergency headquarters should issue a red warning that battery boats are prohibited from sailing
5 BFS	29.63-40.75 km/h	All boats are prohibited from sailing

2.6.2 Safety equipment

Safety equipment on board complies with international standards in place for the floating classroom and recreational boating (Table 2-5).

Table 2-5 Safety equipment

Symbol	Name (requirement)	Quantity
	life jacket (Level 50S or greater)	1 person[*]
	anchor and chain/line (to suit vessel size/weight and sea floor)	1
	bailer/bucket/fire bucket (with lanyard)	1
	bilge pump – manual or power operated	1[**]
	fire extinguisher (vessels with electric start electric motors, battery, gas or fuel stoves)	1[**]
	paddles or oars and rowlocks (in vessels under 6 meters unless a second means of propulsion is fitted)	1
	safe label (appropriate to vessel type)	1
	sound signal (air horn/whistle/bell)	1
	water proof torch (floating and operational)	1

*Life jackets must be suitable for the intended wearer, in good condition
**Additional bilge pumps and fire extinguishers may be required for larger vessels

2.6.3 Safety procedure aboard

The captain and science instructors need have to be trained in safety procedures. In the case of an emergency, the captain is responsible for directing the response effort, which will be carried out by the crew and instructors. His orders must be followed, and everyone must cooperate to the fullest. Due to heightened security concerns, all personnel and baggage are subjected to search. Student backpacks and other non-necessary items are not allowed on the boat. All participants required to wear personal flotation devices during the boat trip. Personal flotation devices (life jackets) are always available for all participants.

1. Before boarding

One of the aquatic science instructor guides will:

(1) Welcome the group and introduce the captain and instructors;

(2) Give a safety lecture covering: off limit areas;personal flotation devices and emergency procedures and additional safety rules.

2. Boarding process

When the group boards the boat, the instructors will assist anyone to wear personal flotation devices. The captain or instructor will point out the various areas of the boat:

(1) Cabin—Location of the storage boxes for observations/measurements and safety equipment and tables for writing and poisoning of instruments like microscopes.

(2) Off limits, i.e. pilot house and roof.

3. After setting sail

During the cruise, an aquatic science instructor will orient the participants to the equipment used on the deck. Most of the above equipment is for sampling. However, the Secchi disk is used to measure water transparency, the Forel-Ule Color Scale is used to determine the color of the water, and nets are used for sampling phytoplankton and other aquatic organism. Binoculars are used for bird watching and microscope for phytoplankton analysis.

2.6.4 Safety management tips for instructors

1. Be certain you know in advance the time and place to meet the vessel. The trip starts at the port of the HLNZ-C.

2. Be prepared for any weather conditions. The vessels will go out regardless of weather conditions with a few exceptions. For example, trips are not canceled on cold, rainy days, but they will be canceled of severe weather conditions, especially storms (Table 2-4).

3. Advise participants to wear appropriate clothing. Appropriate attire is

long pants or shorts depending on the time of year. It can be colder on the lake than inland and it is usually windy as well the group may want to bring jackets. Also, they should wear rubber-soled, closed toe shoes on-board. No open sandals or flip flops! A wet deck is very slippery. Those sensitive to the sun should wear hats, long sleeve shirts, long pants, and should bring the proper sunscreen.

4. Limit the items that are brought on-board.Backpacks are not needed on the cruise and must be left onshore. There is little room for storage. Leave lunches onshore since there will be no opportunity for eating on the vessels.

5. Determine special needs of the group. Let organizer know of any special needs. The current catamaran has limited handicapped accessibility. For those needing medication for motion sickness, take medication at the proper time before going on-board the vessel. If there is a need for special medication or if you are allergic to insects such as bees and wasps, take any needed medication with you while on-board the vessel.

Chapter 3 Learning units of the water science activities

In this booklet, the term learning unit is used deliberately (the correspondence between unit and topic is detailed in Table 3-1). A unit houses a specific topic. A unit, depending on its length and topic, could take one lesson (i.e., one class lecture) to finish, or several lessons, or more.

Table3-1 Overview of learning units

Code	Title	Reference
Learning unit 1	Water depth measuring	3.1
Learning unit 2	Water sampling	3.2
Learning unit 3	Water temperature measuring	3.3
Learning unit 4	Water transparency measuring	3.4
Learning unit 5	Water color measuring	3.5
Learning unit 6	Water turbidity measuring	3.6
Learning unit 7	Water conductivity measuring	3.7
Learning unit 8	Water pH measuring	3.8
Learning unit 9	Water alkalinity measuring	3.9
Learning unit 10	Dissolved oxygen measuring	3.10
Learning unit 11	Lake bottom sampling	3.11
Learning unit 12	Plankton sampling	3.12

General symbols for category of sampling as easy reference see below, a combination is possible as well.

Physical testing   Chemical testing   Biological testing   Data analysing

3.1 Learning unit 1: Water depth measuring

3.1.1 Instructor's guide

Type	Physical testing
Objectives	1. Know some reasons why it is important to know the depth of the water 2. Know how to determine the depth of the water
Aids	special rope with knots and anchor, hydrosphere data recording sheet

3.1.2 How is the depth of the water determined?

There are at least two reasons why it is important to know the depth of the water below the surface:

1. To prevent running the vessel aground.

2. To be able to relate scientific findings to the depth of the water from which samples are taken.

Many water quality parameters such as temperature and dissolved oxygen vary with depth as well as with the time of day. The depth of light penetration, which is influenced by turbidity, has an effect on the productivity of plants and algae in an aquatic ecosystem. Various depths in a lake host different assemblage of benthic (bottom-dwelling) organisms. Plankton and fish move from one depth to another based on changing environmental conditions.

A simple and old fashion method for finding the depth of water is to lower a weight attached to a rope over the side of the vessel. When the weight touches the bottom, the rope becomes slack. The rope is then pulled back on-board and the length of the rope needed to touch the bottom is determined.

3.1.3 Frequently asked questions

How to ensure the quality of measured data?

In order to ensure the quality of the drawing, it is necessary to carefully check and analyze the measured data. For example, the accuracy of plane control, elevation control and positioning points, the completeness of navigational obstacle detection, the rationality of sounding line layout, the density of depth points, and the accuracy of contour contour drawing, etc. The median error of the anchor point should be used to evaluate the accuracy of the anchor point.

3.2 Learning unit 2: Water sampling

3.2.1 Instructor's guide

Type		Physical and chemical testing
Objectives	1. Know how water is sampled on the vessels 2. Explain what Van dorn water sampling bottles are 3. Conduct proper water sampling	
Aids	Van Dorn water sampling bottle, hydrosphere data recording sheet	

3.2.2 How is water sampled on the vessels?

Two kinds of sampling bottle are used on the vessel: Van Dorn water sampling bottle and, less frequently, Kemmerer water sampler. The idea behind these samplers is to collect water at a known depth. The Van Dorn bottle are used for sampling at various depths.

3.2.3 What is Van Dorn water sampling bottle?

The Van Dorn bottle provides a means of obtaining water samples at selected depths below the surface. It consists of an open ended clear plastic cylinder that can be attached to the hydrographic wire (the steel wire wound on the winch) and lowered to any desired depth. A deckhand operates the winch. The bottle also provides a platform to which thermometer can be attached to record the temperature of the water at the location of each Van Dorn bottle.

Each end of the cylinder is fitted with a rubber cover. The open Van Dorn bottle is attached to the line with the covers pulled out and twisted back and around to the side. The bottle is lowered to a pre-selected depth and left there until the thermometer attached inside equilibrate with the water at that depth.

When it is time to lower the Van Dorn bottle into the water (this is called making a "cast"), a decision is made about the depth to which to send the bottle. This decision is based upon the depth of the water at the station and the number of samples needed. Normally only two water samples (surface and bottom) are taken. If a third water sample is required, the Kemmerer bottle may be used to obtain the surface sample. One Van Dorn bottle could then be used to obtain a mid-depth and bottom sample for some trips. In shallow areas, typically only one water sample is taken.

Water samples from each bottle can then be taken for analysis and the temperature read from the attached thermometer. You may be able to see organisms in the water sample through the clear wall of the Van Dorn bottle.

3.2.4 Frequently asked questions

How to store water samples?

The storage methods for water samples include: (1) refrigeration or cryopreservation; (2) storage with chemical reagents. Example: $HgCl_2$ is generally added to water samples for determination of ammonia nitrogen, nitrate nitrogen, and chemical oxygen demand to inhibit biological oxidation and reduction. The water samples for measuring metal ions are generally acidified with HNO_3 to a pH of 1 - 2 to prevent precipitation and adsorption of metal ions.

3.3 Learning unit 3: Water temperature measuring

3.3.1 Instructor's guide

Type	Physical testing	
Objectives	1. Know how water temperature is measured on the vessels 2. Describe what is the significance of temperature data. List at least 3 reasons 3. Investigate the reasons for changing in the temperature of the Hengshui Lake, how the Hengshui Lake temperature varies throughout the year 4. Know how to use a thermometer	
Aids	hydrosphere data recording sheet, latex gloves, clock or watch, enough string to lower the thermometer into the water, alcohol-filled thermometer, distilled water, 500 mL beaker	

3.3.2 How is water temperature measured on the vessel?

Usually water temperature is measured on the vesselusing thermometer that is mounted on the sampling bottle. The thermometer on the sampling bottle must be read immediately after the bottle is retrieved. Temperature is recorded in degrees Celsius (°C).

3.3.3 What is the significance of measuring the temperature of water?

Several interesting studies can be made from the data obtained from the temperature readings. Most prominent is the identification of temperature patters at various depths and the relationship to the seasons of the year.

The temperature readings also give an indication of whether conditions are favorable for cold water fishes. The rates of metabolism of animals, as well as rates of photosynthesis and decomposition, are temperature sensitive. The migration of fish and their spawning behavior are associated with temperature changes. Temperature and dissolved oxygen are related in that warmer water holds less oxygen than cooler water.

3.3.4 How does the lake temperature vary throughout the year?

In most inland lakes in the temperate zone, the temperature of the lake is essentially uniform from top to bottom two times per year, spring and late fall (Figure 3-1).

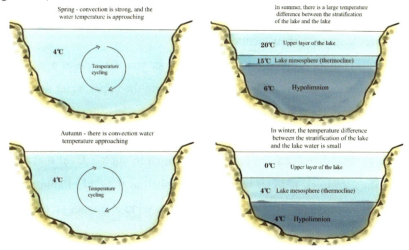

Figure 3-1　Seasonal variation of lake water temperature

During winter, temperate zone lakes generally achieve a relatively uniform temperature from surface to bottom with slightly colder water near the surface. Once covered with ice, the water just beneath the ice is slightly above the freezing point, and increases to no more than 3.98 °C (the density of water is greatest at 3.98 °C and not zero) towards the bottom of the lake. It prevents lakes from freezing from the bottom up to the surface. It is because of the direct relationship between water temperature and density that ice floats on the surface of a lake. The ice layer provides protection from mixing of water by wind, inhibits diffusion of oxygen.

The relationship between water temperature and density also plays a determining role in lake conditions during the summer. Remember that above about 3.98 °C, the water density decreases and this warmer, lighter water floats on colder, heavier water. Hence, as heat from the sun increases during the summer, the upper water of the lake becomes warmer and lighter than the deeper water. The energy of the wind is inadequate to mix the upper water with the colder, more dense, deeper water. This situation leads to a summer stratification period with the warmer water separated from the deeper, much colder, bottom water.In the summer, lakes show thermal stratification, which means that there are distinct layers defined by temperature. When stratification occurs, the different layers are given names to identify their location.

One very important consequence of summer stratification of a lake is that circulation due to wind action is largely confined to the upper water mass known as the epilimnion. Because the lower water mass is isolated from the atmosphere and receives little, if any, sunlight, dissolved oxygen is not replenished in this water mass. The dissolved oxygen may diminish to such a level that it limits aquatic life. However, oxygen will be replenished at the end of summer when water temperatures become more uniform and the wind circulates all of the water in the lake basin.

Hengshui Lake is a shallow lake, with a mean water depth that ranges approximately from 1.5 m to 3 m with some areas of 6 - 8 m depth.

3.3.5 Measuring the temperature of your water using an alcohol-filled thermometer

Aids

◎ hydrosphere data recording sheet   ◎ alcohol-filled thermometer (with rubber band attached)

◎ clock or watch   ◎ latex gloves

Operating flow

1. Fill out the top portion of your hydrosphere data recording sheet.

2. Put on the gloves.

3. Slip the rubber band around your wrist so that the thermometer is not accidentally lost or dropped into the water.

4. Check the alcohol column on your thermometer to make sure there are no air bubbles trapped in the liquid. If the liquid line is separated, notify your teacher.

5. Put the bulb end of the thermometer into the sample water to a depth of 10 cm.

6. Leave the thermometer in the water for three minutes.

7. Read the temperature without removing the bulb of the thermometer from the water.

8. Let the thermometer stay in the water sample for one more minute.

9. Read the temperature again. If the temperature has not changed, go to Step 10. If the temperature has changed since the last reading, repeat Step 8 until the temperature stays the same.

10. Record the temperature on the hydrosphere data recording sheet.

11. Have two other students repeat the measurement with new water samples.

12. Calculate the average of the three measurements.

13. All temperatures should be within 1.0 °C of the average. If they are not, repeat the measurement.

3.3.6 Frequently asked questions

Why is the water temperature sometimes colder and sometimes warmer than the air temperature?

Water has a higher specific heat than air. This means it takes water longer to heat up and longer to cool down than it does air. As a result, air responds much more quickly than water to changes in temperature.

3.4 Learning unit 4: Water transparency measuring

3.4.1 Instructor's guide

Type	Physical testing	
Objectives	1. Define water transparency and explain how environmental variables 2. Identify a Secchi Disk and explain what the significance of Secchi disk readings is 3. Conduct transparency measurements in the field with Secchi disk properly	
Aids	hydrosphere data recording sheet, latex gloves, Secchi disk transparency protocol field guide, Secchi disk (with rope), meter stick, clothespins (optional)	

3.4.2 What is water transparency?

The transparency of water relates to the depth that light will penetrate water. The transmission of light into a body of water is extremely important since the sun is the primary source of energy for all biological phenomena. Light is necessary for photosynthesis, a process that produces oxygen and food for consumers.

It is common practice to consider the depth of the euphotic zone (the upper layers of a body of water into which sufficient light penetrates to permit growth of green plants) to be 2.7 times (roughly 3 times) the limit of visibility.

Water clarity is related to amounts of suspended particles as well as amounts of phytoplankton and zooplankton.

3.4.3 What is a Secchi disk?

The Secchi disk is a very simple device about 20 cm in diameter made of metal or plastic (Figure 3-2).

The Secchi disk provides a means for determining the limit of visibility that is based on contrast. The upper surface of the Secchi disk is divided into four quadrants that are alternatively black or white. An eyebolt is located at the center of the disk on the upper side so that a line can be tied to the disk. This makes it possible to lower the disk into the water from a boat or dock. A weight is attached to the underside of the disk so that the equipment will sink below the surface. This line is marked every 0.5 meter making it possible to determine the depth at which the Secchi disk disappears from sight as it is lowered into the water.

Figure 3-2 Secchi disk

The Secchi readings are a semi-quantitative measure of water transparency since a variety of factors such as the time of day, sky and water surface conditions, and differences between observers will give varying depths for the same location. It is possible that each person will have a different opinion of the depth at which the disk disappears from sight. That is why it is important that Secchi disk records contain information about the conditions under which the readings were taken.

Standard conditions for Secchi disk measurements include a clear sky, sun directly overhead, and minimal waves or ripples. These measurements must be taken on the shaded and protected side of the vessel. Any deviations from these conditions should be clearly stated in the data. It is interesting to note that visibility in water is roughly twice the Secchi depth since the light must travel twice through a column of water equal in length to the Secchi depth from the surface to the disk and then back up again after being reflected from the disk.

3.4.4 What is the significance of Secchi disk readings?

Secchi disks are standard tools for inland lake monitoring. Volunteer groups throughout the different states and countries take Secchi disk readings to indicate the current status of their lake and to compare with data from previous years.

In oligotrophic or low nutrient lakes are often greater than 5 meters, whereas eutrophic or nutrient rich lakes have readings less than 2.5 meters.

3.4.5 Measuring the transparency of your water using a Secchi disk

Aids

◎ hydrosphere data recording sheet ◎ meter stick

◎ latex gloves ◎ clothespins (optional)

◎ Secchi disk with rope attached

Operating flow

1. Fill in the top portion of the hydrosphere data recording sheet.

2. Record the cloud and contrail types and cover.

3. Stand so that the Secchi disk will be shaded or use an umbrella or piece of cardboard to shade the area where the measurement will be made.

4. Establish a reference height. This can be a railing, a person's hip, or the edge of a dock. All measurements should be taken from this point. Wear latex gloves, as you will probably touch the rope wet with sample water.

5. Lower the disk slowly into the water until it just disappears.

6. Mark the rope with a clothespin at the water surface (Figure 3-3 a), if you cannot easily reach the water surface (for example, if you are standing on a dock or bridge), mark the rope at your reference height.

7. Lower the disk another 10 cm into the water, then raise the disk until it reappears (Figure 3-3 b).

(a) Lower the disk into the water until it just disappears. Mark the rope with a clothespin at the water surface

(b) Raise the disk until it reappears. Mark the rope with a clothespin at the water surface

Figure 3-3 The use of the Secchi disc

8. Mark the rope with a clothespin at the water surface or at your reference height.

9. There should now be two points marked on the rope. Record the length of the rope between each mark and the Secchi disk on your hydrosphere data recording sheet to the nearest cm. If the depths differ by more than 10 cm, repeat the measurement and record the new measurements on your hydrosphere data recording sheet.

10. If you marked the rope at the water surface, record "0" as the distance between the observer and the water surface.

11. If you marked the rope at a reference point, lower the disk until it

reaches the surface of the water and mark the rope at the reference point. Record the length of the rope between the mark and the Secchi disk as the distance between the observer and the water surface. Repeat steps 5 - 11 two more times with different students observing.

3.5 Learning unit 5: Water color measuring

3.5.1 Instructor's guide

Type		Physical testing
Objectives	1. List what influences the color of the water 2. Describe what is the Forel-Ule color scale 3. Explain what the significance of the color of the water is 4. Know how to use and interpret Forel-Ule color scale	
Aids	Forel-Ule color scale	

3.5.2 What influences the color of the water?

The description of the color of a body of water varies from one person to the next. Many variables affect our perception of its color. Among them are sky conditions, time of day, surface conditions, suspended materials, and the direction from which the water is viewed.

Sky conditions include the presence or lack of clouds. Water viewed on a clear sunny day will appear blue unless there is a sufficient quantity of suspended or dissolved material in the water to affect its color. When the same water is observed on a cloudy day, it will look gray. If there is a large amount of water vapor or dust in the air, the color of the water will be affected.

The altitude of the sun will change during the day. When the sun is seen near the horizon at sunrise or sunset, the color of the water will appear very dark. Later in the day, the water will appear blue or bluish green in the absence of suspended material.

A flat water surface viewed with reflected light can appear to be without color while the same water will take on a color if the surface is covered by waves. If the waves are large and foam from white caps is present, the apparent color of the water can be affected by light passing through the waves.

There are 3 main natural components, besides the water itself, affecting the water color and clarity in natural environments:

(1) Phytoplankton, which is generally colored green;

(2) Dissolved organic matter, which are colored from yellow to brownish;

(3) Other matter, such as fine soils, are colored ranging from milky, grey to dark.

3.5.3 What is the Forel-Ule color scale?

Since there are many variables that affect the apparent color of the body of water including the perception of the person making the observation, it is necessary to establish a standard method for determining color. One technique uses a series of colors obtained by using standardized chemical procedures for producing colored water. Each of these colored samples is contained in a sealed glass tube. The colors range from blue to a range of greens and yellow-browns, and are composed of various solutions of potassium chromate, cobaltous sulfate, and cupric sulfate. These colors form the Forel-Ule color scale. Each color is given a number. Number 1 is a pale blue and number 22 is brown (Figure 3-4).

A determination of water color is made by lowering the Secchi disk from the shady side of the vessel into the water to a depth of one meter. The observer then compares the color of the standard samples in the tubes to the color of the water as seen against the white portion of the Secchi disk.

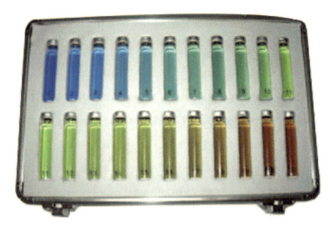

Figure 3-4 Forel-Ule color scale

The apparent color of the water as determined by the above technique is very subjective. Two observers viewing the same water can disagree within one or two numbers. However, if the same technique is used each time color is to be determined, usable data can be obtained.

A trusted method for measuring water quality through color comparison. NOTE: The Forel-Ule colors are contained in glass vials that are subject to being easily broken. Care should be taken when they are used to guard against breakage.

3.5.4 What is the significance of measuring the color of water?

The color of the water can give an indication of its quality and composition.

Color can be associated with material dissolved or suspended within it. The altitude of the sun, water vapor in the air, clouds, and dust also relate to water color (Figure 3-5).

Figure 3-5 Color of natural waters

Deep blue may indicate low amounts of organic matter or plankton.

Green shades are associated with phytoplankton and high biological productivity.

Brown is indicative of mineral matter, organic matter, or large populations of diatoms.

The water in the Hengshui Lake is green colored (Figure 3-6) because of the breakdown of natural organic matter in the wetlands. This matter includes tannins, lignins, and humic acids that are a result of decay of plants and wood.

River contains organic matter in addition to suspended clay minerals. Turbidity measurements can help differentiate between dissolved and suspended materials that color the water.

Figure 3-6 Sample of water color of Hengshui Lake

3.5.5 Measuring the color of your water using the Forel-Ule color scale

1. On the shady side of the vessel, a second person should lower the Secchi disk to a depth of one meter.

2. The observer then compares the Forel-Ule color scale to the color of water as seen against the white portion of the Secchi disk. Record the Forel-Ule color number that is the closest match.

3. Repeat the observations by having two other people determine the color match.

4. Interpret the result.

3.5.6 Frequently asked questions

The color of water is commonly distinguished by surface color and true color. What are the meanings?

Water color is the appearance index of water quality, and is divided into surface color and true color. The true color refers to the color of the water after the suspended particles is removed. The color of the water with suspended particles is called the surface color. For clean or very low turbidity water, the true color is similar to the surface color. For the industrial wastewater and sewage with deep coloration, the true color and the surface color are quite different.

3.6 Learning unit 6: Water turbidity measuring

3.6.1 Instructor's guide

Type	Physical testing	
Objectives	1. Describe what is turbidity 2. Explain how turbidity measured is 3. Explain what the significance of turbidity is 4. Apply currently a turbidity tube	
Aids	hydrosphere data recording sheet, 100 mL beaker, latex gloves, sample bucket, turbidity tube	

Another way to study water clarity is to use a turbidity meter.

3.6.2 What is turbidity?

Turbidity, or cloudiness, in water is caused by a variety of suspended materials. The material can be both organic (plankton, sewage) and inorganic (silt, clay). The suspended material will scatter and absorb light passing through the water. The light scattered back to the observer can be affected so that the water will have a color dependent upon the type and amount of suspended matter. The cloudiness and color can be observed if a sample of water in a transparent container is held between the observer's eye and a light source. It is this phenomenon that is used in the turbidity meter.

3.6.3 How is water turbidity measured?

A nephelometer or spectrophotometer (turbidity meter) measures the cloudiness or opaqueness of a water sample. A nephelometer contains a source of light, a photocell, and a meter. Light is beamed through a water sample. The path of the light is 90 degrees to the direction in which the photocell points. When a sample is placed in the light beam, light scattered by the suspended material in the sample is detected by the photocell. The photocell converts the scattered light into an electrical current that is sent through the meter. The position of the needle on the meter or a digital read-out gives an indication of the turbidity of the water sample.

Turbidity can be determined with the nephelometric method from the root meaning "cloudiness". This word is used to form the name of the unit of turbidity, the NTU. This acronym stands for nephelometrie turbidity unit.

The meter reading cannot be used to compare the turbidity of different water samples unless the instrument is calibrated. The science instructors calibrate the meter regularly. Calibration consists of adjusting the meter reading to a known value when a standard sample is placed in the light beam. A standard suspension is often made from a polymer called Formazin, which has stable reproducibility.

3.6.4 What is the significance of measuring the turbidity of water?

Turbidity relates to the effect that suspended particles have on water clarity. High turbidity readings (low clarity) can indicate erosion and sedimentation problems. Rainfall and runoff can increase the suspended solid load in a river and make the river appear cloudy or muddy. High biological productivity related to increases in nutrients and temperature can result in increases of diatoms and other algae that contribute to turbidity. Turbidity meters can be used to estimate plankton density.

Turbidity readings in Hengshui Lake are likely to range from 15.7 to 61.2

NTU (Table 3-2). The Yellow River often ranges from 20 to 11 000 NTU.

Table 3-2 The turbidity in Hengshui Lake from April to November (NTU)

Date	Open zone	Reed zone	Typha zone	Small lake
April	26.5	—	—	30.0
May	20.6	15.7	22.4	41.4
June	33.5	21.8	30.2	27.5
July	34.1	28.7	31.4	38.9
August	48.8	26.1	31.5	—
September	61.2	31.1	40.8	54.1
October	51.1	—	—	53.9
November	24.6	—	—	34.6

Elevated turbidity can cause an increase in temperature since suspended particles absorb heat. Reduction of light penetrating the water column due to turbidity can decrease the rate of photosynthesis. This, in turn, can decrease the amount of dissolved oxygen in the water. As suspended particles settle, they can impair the habitat needed for fish spawning and aquatic macroinvertebrates. They can also clog the gills of fish and the breathing apparatus of inverte-brates. Particles serve as places of attachment for harmful microorganisms and toxic materials. Turbidity in drinking water is decreased through the process of flocculation, which involves addition of alum or a mixture of iron, lime, and chloride to cause solids to settle out.

3.6.5 Instructions for use of a turbidity tube

1. Pour water drawn in a bucket into the tube until the black and white image at the bottom of the tube is no longer visible when looking directly through the water column (Figure 3-7).

2. Rotate the tube while looking down at the image to see if the black and white areas of the decal

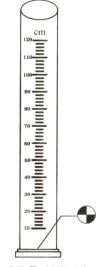

Figure 3-7 Turbidity tube

are distinguishable.

    3. Record this depth of water to the nearest 1 cm.

    4. Enter data for each observer, and calculate the average of the different observations. If you can still see the image on the bottom of the tube after filling it, simply record the depth as greater than (>) the depth of the tube.

3.6.6 Measuring the turbidity of your water using a turbidity tube

Aids

◎ hydrosphere data recording sheet   ◎ 100 mL beaker

◎ sample bucket   ◎ latex gloves

◎ turbidity tube

Operating flow

    1. Fill in the top portion of the hydrosphere data recording sheet.

    2. Record the cloud types, contrail types and cover.

    3. Put on gloves.

    4. Collect a surface water sample.

    5. Stand with your back to the sun so that the turbidity tube is shaded.

    6. Pour sample water slowly into the tube using the 100 mL beaker. Look straight down into the tube with your eye close to the tube opening. Stop adding water when you cannot see the pattern at the bottom of the tube.

    7. Rotate the tube slowly as you look to make sure you cannot see any of the pattern.

    8. Record the depth of water in the tube on your hydrosphere data recording sheet on the nearest cm. Note: If you can still see the disk on the bottom of the tube after the tube is filled, record the depth as " >120 cm".

    9. Pour the water from the tube back into the sample bucket or mix up the remaining sample.

    10. Repeat the measurement two more times with different observers using the same sample water.

3.6.7 Frequently asked questions

What are the differences and connections among turbidity, color and transparency of water?

Turbidity is the degree of obstruction of light transmission by suspended matter in water. Due to the presence of insoluble substances in the water, part of the light passing through the water sample is absorbed or scattered, but not linearly penetrated.Turbidity and color are both optical properties of water, but they are different. The water color is caused by dissolved substances in water, while turbidity is caused by insoluble substances in water. Therefore, some water samples have high chroma but not turbid, and vice versa.Transparency refers to the degree of clarity of the water sample, and clean water is transparent. The more suspended solids and colloidal particles in the water, the lower the transparency. Generally, the transparency of groundwater is relatively high. Transparency is an indicator of water quality related to the combined effects of both water color and turbidity.

3.7 Learning unit 7: Water conductivity measuring

3.7.1 Instructor's guide

Type	Chemical testing
Objectives	1. Describe what is conductivity 2. Describe how is conductivity measured 3. Explain what the significance of conductivity is 4. Use an electrical conductivity meter 5. Examine reasons for changes in the electrical conductivity of a water body

(To be continued)

(Continued)

Type	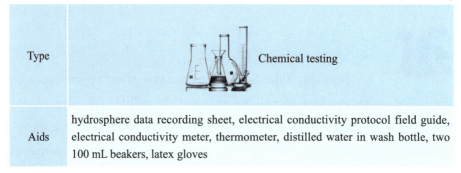 Chemical testing
Aids	hydrosphere data recording sheet, electrical conductivity protocol field guide, electrical conductivity meter, thermometer, distilled water in wash bottle, two 100 mL beakers, latex gloves

3.7.2 What is conductivity?

Conductivity or specific conductance is the measure of the water's ability to conduct an electrical current. Conductivity depends upon the number of ions or charged particles in the water. The ease or difficulty of the flow of electrical current through liquids makes it possible to divide them into two broad categories: electrolytes and nonelectrolytes. Electricity passes easily through water that is high in electrolytes or ions, and poorly through low electrolyte materials such as pure water or many organic solvents such as alcohol or oil.

3.7.3 How is conductivity measured?

A conductivity meter is used to measure the ability of the water sample to conduct electricity. The specific conductance is measured by passing a current between two electrodes that are placed into a sample of water. The unit of measurement for conductivity is expressed in either micro Siemens (μS/cm).

The warmer the water, the higher the conductivity with an increase of about 1.9% per Celsius degree. Conductivity is reported at standard temperature of 25.0 °C. The Range of conductivity for different categories of water are shown in Table 3-3.

Table 3-3 Range of conductivity for different categories of water

Category of water	µS/cm
Distilled water	0.5 - 2
Drinking water	50 - 1 500
Wastewater	>10 000

3.7.4 What is the significance of measuring the conductivity of water?

Conductivity determinations are useful in aquatic studies because they provide an estimate of dissolved ionic matter in the water.

Low values of specific conductance are characteristic of high-quality, oligotrophic (low nutrient) lake waters.

High values of specific conductance are observed in eutrophic lakes where plant nutrients (fertilizer) are in greater abundance.

Very high values are good indicators of possible pollution sites. For instance, industrial discharges, road salt, and failing septic tanks can raise conductivity.

A sudden change in conductivity can indicate a direct discharge or other source of pollution into the water.

3.7.5 Frequently asked questions

What is the relationship between the conductivity and salinity of water?

The salinity in the water sample is ionized into an ionic state and has conductivity. The higher the ion concentration (that is the higher salinity), the higher the conductivity, and vice versa.

3.8 Learning unit 8: Water pH measuring

3.8.1 Instructor's guide

Type	Chemical testing
Objectives	1. Describe what is pH,understand the differences among acid, basic and neutral pH values 2. Explain how pH is measured 3. Explain what is the significance of pH 4. Use a pH meter or pH paper 5. Examine reasons for changes in pH of the Hengshui Lake
Aids	**For measuring pH with pH paper:** hydrosphere data recording sheet, using pH paper (electrical conductivity greater than 200 μS/cm) field guide or using pH paper (electrical conductivity less than 200 μS/cm) field guide, pH paper, 50mL or 100 mL beaker, latex gloves **For measuring pH with the pH meter:** hydrosphere investigation data sheet, using a pH meter (electrical conductivity greater than 200 μS/cm) field guide or using a pH meter (electrical conductivity less than 200 μS/cm) field guide, pH meter, distilled water, clean paper towel or soft tissue

3.8.2 What is pH?

A natural body of water can be acidic, neutral, or basic. Many factors determine this condition including the composition of the material forming the basin holding the water, acidity of rain falling into the water, and the condition of water flowing into the body of water from streams, rivers, or storm runoff. The standard measurement used to indicate acidic or basic conditions is called pH with "p" referring to the "power" (puissance) of the hydrogen ion activity.

The pH scale is a series of numbers ranging from 0 to 14 which denote various degrees of acidity or alkalinity. Values below 7 and approaching 0 indicate increasing acidity. Values from 7 to 14 indicate increasing alkalinity. Since the scale is loga-rithmic, the difference between pH 5 and pH 6 is not one but rather ten, that is, pH 5 is ten times more acidic than pH 6.

3.8.3 How is pH measured?

There are several ways to measure pH, which include pH paper, pH pen, and pH meter.

For pH paper, strips of paper are saturated with an indicator that changes color with varying degrees of acidity. The color of the paper is compared to a color scale that is specific to the range and type of paper used. This means of determining pH usually measures only to about 1 pH unit; however, it is inexpensive.

A pH pen is basically a simple electrode similar to that found in a pH meter. Both measure electrical potential associated with the hydrogen ion activity across an electrode immersed in the water sample. Accuracy ranges from 0.1 to 0.01 pH units.

Basic pH meter will have a device to measure voltage, a glass electrode to immerse in the water, a reference electrode that provides a constant electric potential, and a temperature compensation device. The pH readings are temperature dependent. The results are given in either pH units or millivolts (mV).

Paper vs. Meter: Which instrument should you use?

pH paper

Advantages: Easy for young children to use and does not need calibration.

Disadvantages: Resolution is not as good as meter and it is not temperature compensated.

pH meter

Advantages: Measures to 0.1 pH units; may be temperature compensated.

Disadvantages: The meter must be calibrated with buffer solutions before each use, more expensive than pH paper, performance deteriorates over time.

Specific instructions for the model carried on-board are posted next to the instrument. Before it is used, the science instructors calibrate the pH meter.

3.8.4 What is the significance of measuring the pH of water?

Changes on pH can be associated with wastewater discharges and sources of pollution. However, natural changes in pH occur with variations in levels of carbon dioxide. Carbon dioxide is very soluble in water. It enters the water from the atmosphere and is also generated from animal and plant respiration and decomposition. Dissolved carbon dioxide can combine with water to yield carbonic acid. Plants reduce amounts of carbon dioxide through photosynthesis making surface waters more basic.

Water quality standards generally call for a pH between 6.0 to 9.0. A pH between 6.7 and 8.6 will support a well-balanced fish population. Only a very few species can tolerate pH values less than 5.0 or greater than 9.0 (Table 3-4).

For the Hengshui Lake a water samples typically have a pH range of 7.7 - 8.9.

Table 3-4 Effects of pH on fish and algae

Min (pH)	Max (pH)	Effects
4.0	10.1	Limits for the most resistant fish species
5.0	9.0	Tolerable range for the most fish
4.5	9.0	Trout eggs and larvae develop normally
4.6	9.5	Limits for perch
4.1	9.5	Limits for trout
—	8.7	Upper limit for good fishing waters
5.4	11.4	Fish avoided waters beyond these limits
6.0	7.2	Optimum (best) range for the fish eggs
7.5	8.4	Best range for the growth of algae

3.8.5 Measure the pH of your water using pH paper

Aids

◎ hydrosphere data recording sheet   ◎ latex gloves

◎ pH paper (electrical conductivity greater than200 μS/cm)

◎ 100 mL beaker

Operating flow

1. Fill in the top part of your hydrosphere data recording sheet.

2. In the pH section of the hydrosphere data recording sheet, check the box next to "pH paper".

3. Put on latex gloves.

4. Rinse the beaker with sample water three times.

5. Fill the beaker halfway with sample water.

6. Follow the instructions that come with your paper for testing the pH of the sample.

7. Record your pH on the hydrosphere data recording sheet as Observer 1.

8. Repeat steps 4 - 6 using new water samples and new pieces of paper. Record the data on the hydrosphere data recording sheet as Observer 2 and Observer 3.

9. Find the average of the three observations.

10. Check to make sure that each observation is within 1.0 pH units of the average. If they are not within 1.0 units of the average, repeat the measurements. If your measurements are still not within 1.0 pH units of the average, discuss possible problems with your teacher.

11. Discard used pH paper and gloves in a waste container. Rinse the beaker with distilled water.

3.8.6 Frequently asked questions

1. Why could I not find a color match with the pH paper?

The conductivity of your water might be low (see Electrical Conductivity Protocol). The pH paper takes longer to react with the water if the conductivity is less than 400 micro-Siemens/cm (μS/cm). If your water has a conductivity of less than 300 μS/cm, some pH paper does not work well. Another reason you may have problems is if your pH paper is old or has not been stored properly.

2. What do I do if the pH seems to be between two colors matches on the box?

Report the match that is the closest. This is the reason we have three students do the protocol. Taking the average of the three readings gives a more accurate measurement.

3.9 Learning unit 9: Water alkalinity measuring

3.9.1 Instructor's guide

Type		Chemical testing
Objectives	1. Describe what is alkalinity 2. Explain how alkalinity is measured 3. Explain what the significance of alkalinity is 4. Use an alkalinity kit 5. Examine reasons for changes in the alkalinity of a water body 6. Explain the difference between pH and alkalinity	
Aids	alkalinity test kit, hydrosphere data recording sheet, soda, alkalinity protocol field guide, distilled water, latex gloves and safety goggles	

3.9.2 What is alkalinity?

Alkalinity is a measure of the capacity of water to neutralize acids. This is known as the acid neutralizing (buffering) capacity of water or the ability of water to resist a decrease in pH when acid is added. Alkalinity in water is due primarily to the presence of bicarbonate (HCO_3^-), carbonate (CO_3^{2-}), and hydroxide ions (OH^-). It relates to the balance of carbon dioxide in water and is a function of pH. The HCO_3^-, CO_3^{2-}, CO_2 equilibrium system accounts for the major buffering mechanism in water.

3.9.3 How is alkalinity measured?

Alkalinity is expressed as phenolphthalein alkalinity or total alkalinity. Both types can be determined by a titration with standard sulfuric acid to an endpoint pH. Indicators such as phenolphthalein and bromcresol green-methyl red define endpoints or a pH meter could be used for determination of endpoints.

Phenolphthalein alkalinity is determined by titration to a pH of 8.3 and indicates the total hydroxide and half of the carbonate present.

Total alkalinity is determined by titration to a pH of 5.1, 4.8, 4.5 or 3.7 depending upon the amount of carbon dioxide present. Generally, 4.5 is used.

The unit of measurement for alkalinity is usually mg/L. Another measurement unit for alkalinity is milliequivalents per liter.

3.9.4 What is the significance of measuring the alkalinity of water?

Alkalinity is the measure of the resistance of water to the lowering of pH when acids are added to the water. Acid additions generally come from rain or snow, though soil sources are also important in some areas. Alkalinity increases as water dissolves rocks containing calcium carbonate. Carbonates and hydroxide may be significant when algal activity is high and in industrial water.

When a lake has too little alkalinity, typically below about 100 mg/L, a large influx of acids from an intense rainfall or rapid snowmelt event could (at least temporarily) consume all of the alkalinity. This results in a drop of the pH of the water to levels harmful for amphibians, fish or zooplankton. Lakes and streams in areas with little soil, such as in mountainous areas, are often low in alkalinity. Because pollutants tend to wash out of a snow pack during the first part of snowmelt, there is often a higher influx of acidic pollutants in spring, which is also a critical time for the growth of aquatic life.

High alkalinity can mitigate metal toxicity by using available bicarbonates and carbonates to take metal out of solution. The metals would thus be unavailable to fish and other aquatic organisms.

3.9.5 Measure the alkalinity of your water using a test kit

Aids

◎ hydrosphere data recording sheet　　◎ distilled water

◎ alkalinity test kit　　　　　　　　◎ safety goggles

◎ latex gloves

Operating flow

1. Fill out the top portion of your hydrosphere data recording sheet.

2. Put on the gloves and goggles.

3. Follow the instructions in your alkalinity test kit to measure the alkalinity of your water.

4. Record your measurement on the hydrosphere data recording sheet as Observer 1.

5. Repeat the measurement using fresh water samples.

6. Record as Observers 2 and 3.

7. Calculate the average of the three measurements.

8. Each of your individual measurements should be within the acceptable range of the average.

9. If one measurement is outside this range, discard that measurement and find the average of the other two.

10. If they are still in range, report only the two measurements.

11. If more than two of your measurements are not in range, repeat from Step 3.

3.9.6 Frequently asked questions

What is the difference between T alkalinity, M alkalinity and P alkalinity?

T alkalinity is total alkalinity, which is a measure of the total amount of substances in the sample that can be neutralized by H^+. M alkalinity is methyl orange alkalinity, which is the alkalinity measured when methyl orange is used as an indicator. P alkalinity is Phenolphthalein alkalinity, which is the alkalinity measured when phenolphthalein is used as an indicator. When methyl orange is used as an indicator, the discoloration range is 3.1 - 4.4, and when phenolphthalein is used as an indicator, the discoloration range is 8 - 10. Therefore, the alkalinity of methyl orange is greater than that of phenol-phthalein. When measuring total alkalinity, methyl orange is used as an indicator. Therefore, methyl orange alkalinity can be regarded as total alkalinity. Taking the measurement of natural water samples as an example, P alkalinity actually measures the content of and OH^- ions, while M alkalinity measures the content of ions in addition to the abovementioned ions.

3.10 Learning unit 10: Dissolved oxygen measuring

3.10.1 Instructor's guide

Type		Chemical testing
Objectives	1. Describe what is dissolved oxygen 2. Explain how dissolved oxygen is measured 3. Describe what is the significance of dissolved oxygen 4. To measure the amount of oxygen dissolved using the DO protocol	
Aids	dissolved oxygen kit or probe, latex gloves, safety goggles, waste bottle with cap, distilled water, thermometer, 100 mL graduated cylinder, 250 mL polyethylene bottle with lid, clock or watch, solubility of oxygen table, correction for elevation table, hydrosphere data recording sheet, oxygen kits data sheet	

3.10.2 What is dissolved oxygen?

Oxygen gas dissolves freely in fresh water.

Increase: Oxygen from the atmosphere as well as that produced as a by-product of photosynthesis may increase the dissolved oxygen concentration in water.

Decreased: Oxygen is removed from the water through the processes of respiration by plants, algae, and animals, as well as the microbes responsible for the decomposition of organic wastes entering the water.

The distribution of dissolved oxygen (DO) within an aquatic environment varies horizontally, vertically, and with time. Its distribution is dependent upon atmospheric contact, wave and current actions, thermal phenomena, waste inputs, biological activity, and other characteristics of a lake or stream.

High levels of oxygen are likely in surface water on windy days. DO levels are temperature and pressure dependent. Cold water has the capacity to hold more oxygen than warm water.

Photosynthesis contributes to an increase in dissolved oxygen levels during the day. However, there are biological processes in water that consume oxygen such as respiration by organisms and decomposition of organic matter by microorganisms. The oxygen consumed by these processes is called the biological oxygen demand or BOD.

When demand for oxygen is high and oxygen production from photosynthesis is not occurring such as before sunrise, dissolved oxygen readings can be low. Dissolved oxygen in the littoral zone may show a 4 - 6 mg/L diurnal fluctuation. In summer, deep areas of a lake would be expected to yield low dissolved oxygen readings.

3.10.3 How is dissolved oxygen measured?

Since an adequate supply of oxygen is necessary to support life in a body of water, a determination of the amount of oxygen provides a means of assessing the quality of the water with respect to sustaining life. There are two chemical methods to determine the amount of oxygen dissolved in a water sample in use:

Winkler Method: Precisely measured amounts of chemicals (reagents) are added to a water sample until a color change is achieved. A color change (or electrical measurement for other types of titration) marks the endpoint of the test.

Dissolved oxygen meter or probe: Units for measuring dissolved oxygen are milligrams per liter (mg/L).

Because the solubility of oxygen in water is dependent upon temperature, pressure, and ionic concentrations, it is also important to calculate percentage saturation. The accompanying Nomogram will permit you to quickly approximate oxygen saturation values (Figure 3-8). The saturation point indicates the level at which water will not generally hold any more oxygen at a given temperature. Supersaturation occurs when the water holds more oxygen molecules than usual for a given temperature. Sunny days with lots of photosynthesis or turbulent water conditions can lead to supersaturation. A water sample is "saturated" at 100% and "supersaturated" above 100%.

As shown in Figure 3-8, hold a ruler or a dark thread to join the observed temperature on the upper scale with the dissolved oxygen reading on the bottom scale. The percentage saturation is read where the ruler or thread intersects the middle scale.

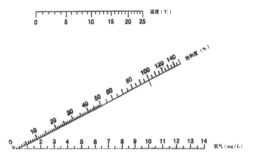

Figure 3-8 Nomogram for dissolved oxygen saturation

3.10.4 What is the significance of measuring the dissolved oxygen of water?

Dissolved oxygen levels provide information about the biological, biochemical, and inorganic chemical reactions occurring in aquatic environments. Most aquatic organisms are highly dependent upon dissolved oxygen and will experience stress, or perhaps even be eliminated from a system, when dis-solved oxygen levels fall below about 3 mg/L. Carp can live in water containing as little as 2 mg/L oxygen.

Poor water quality is also indicated by low percent saturation readings. Levels below 60% may happen with rapid biological processes such as decomposition or high temperatures. Supersaturation can be a problem for organisms in that blood oxygen levels can increase resulting in gas bubbles in the blood.

A general guideline for interpretation of dissolved oxygen readings is:

0 - 2 mg/L: not enough oxygen to support life;

2 - 4 mg/L: only a few kinds of fish and insects can survive;

4 - 7 mg/L: acceptable for warm water fish;

7 - 11 mg/L: very good for most stream fish including cold water fish.

For percent saturation:

Below 60%: poor, water too warm or bacteria using up DO;

60% - 79%: acceptable for most aquatic organisms;

80% - 120%: excellent for most aquatic organisms;

120% or more: too high, may be dangerous to fish.

3.10.5 Measure the dissolved oxygen of your water using a test kit

Aids

◎ hydrosphere data recording sheet   ◎ distilled water

◎ latex gloves

◎ goggles

◎ waste bottle with cap for used chemicals

◎ dissolved oxygen kit

Operating flow

1. Fill in the top of the hydrosphere data recording sheet.

2. Put on the gloves and goggles.

3. Rinse the sample bottle and your hands with sample water three times.

4. Place the cap on the empty sample bottle.

5. Submerge the sample bottle in the sample water.

6. Remove the cap and let the bottle fill with water. Move the bottle gently or tap it to get rid of air bubbles.

7. Put the cap on the bottle while it is still under the water.

8. Remove the sample bottle from the water. Turn the bottle upside down to check for air bubbles. If you see air bubbles, discard this sample. Collect another sample.

9. Follow the directions in your dissolved oxygen kit to test your water sample.

10. Record the dissolved oxygen in your water sample on the hydrosphere data recording sheet as Observer 1.

11. Have two other students repeat the measurement using a new water sample each time.

12. Record their data on the hydrosphere data recording sheet as Observers 2 and 3.

13. Calculate the average of the three measurements.

14. Each of the three measurements should be within 1 mg/L of the average. If one of the measurements is not within 1 mg/L of the average, find the average of the other two measurements. If both of these measurements are within 1 mg/L of the new average, record this average.

15. Discard all used chemicals into the waste container. Clean your dissolved oxygen kit with distilled water.

3.10.6 Frequently asked questions

Why is the determination of dissolved oxygen in the water sample completed in situ?

The dissolved oxygen in the water sample may be partly volatilized into the air due to changes in the external temperature and the length of the storage time. The oxygen molecules in the air may dissolve into the water, which affects the accuracy of the measurement result. Thus, the faster the measurement, the better.

3.11 Learning unit 11: Lake bottom sampling

3.11.1 Instructor's guide

Type	Physical and biological testing
Objectives	1. Know how materials on lake bottom are sampled 2. Identify and sample with a Ponar grab sampler 3. Identify roughly the texture of the bottom materials 4. Understand how the bottom material is studied, and what can be studied already on board 5. Identify several organisms found in the bottom materials 6. Explain the connection between sediments and nutrients
Aids	Ponar grab sampler, benthos and sediment data sheet

3.11.2 How are materials on lake bottoms sampled?

In order to obtain a rough analysis of the bottom materials in water, a variety of devices have been invented. Among them are grab sampler, dredge, corer, and drill. The Ponar grab sampler is the main bottom sampling device used on the vessels to study the composition of the bottom of a lake.

The grab sampler provides a means to obtain a somewhat quantitative and undisturbed sample of the bottom materials. It takes a bite of known surface area and penetration depth, providing the bottom materials is neither too hard or too soft. It is called a grab sampler because of the manner in which it obtains samples.

3.11.3 What is a Ponar grab sampler?

The Ponar grab sampler consists of two opposing semi-circular jaws that are normally held open by a trigger mechanism. The sampler is lowered to the bottom where contact with the bottom sets off the trigger and a strong spring snaps the jaws shut trapping a sample of the bottom inside. Fine copper screen covers the top of the jaws so that the trapped material will not wash out as the sampler is retrieved (Figure 3-9).

Figure 3-9 The Ponar grab sampler

The deckhand normally places the Ponar grab sampler on deck at the start of the cruise. It is placed in an out of the way location until a sample of the bottom materials is desired. Samples are taken while the vessel is on station and not moving through the water.

When a sample is to be taken, the Ponar grab sampler is taken to the platform where the deckhand attaches it to the hydrographic wire (winch line). The sampler is "cocked", that is, the jaws are opened and the trigger is set. The sampler is then swung over the side and lowered to the bottom. The jaws snap shut upon reaching the bottom and a sample of materials is obtained. As long as the Ponar sampler is hanging freely from the hydrographic wire, the trigger mechanism will keep the jaws open. However, as soon as there is slack in the winch line, the trigger will be released. When the winch starts to raise the Ponar grab sampler, the jaws will close, taking a "bite" (sample) from the bottom of the lake. Sometimes when lake waters are rough, the rocking action of the vessel may cause the winch line to become slack enough to release the trigger prematurely, allowing the jaws to close before the sampler has reached the bottom. In such instances the Ponar grab sampler must be brought back aboard the vessel to reset the trigger and a second sampling attempt is carried out.

When a successful Ponar grab sample is brought aboard, the sampler is lowered into a rectangular stainless steel box that has a very fine screen on the bottom side. The deckhand will empty the contents of the Ponar grab sampler into the stainless box and rinse the grab sampler with a hose to make sure that all of the sample is rinsed into the stainless steel box. The bottom sample is now ready for examination.

NOTE: The Ponar grab sampler is a piece of equipment that is operated by the deckhand. It is very heavy. Please stay out of the way when it is being used.

3.11.4 How is bottom materials studied?

The materials brought up from the bottom can be examined in several ways.

A quick visual inspection can give a qualitative description of the kind of materials retrieved: sand, silt, clay, mud, decayed organic, or a combination. In many cases, the sample will reveal the presence of small animals. These can be found by washing the fine sediments through the fine mesh screen and leaving the organisms on the screen where they can be picked from the screen and placed in a plastic Petri dish. When all of the organisms have been collected in the Petri dish, the dish can be taken into the main cabin and examined under the stereo microscope.

The composition of bottom sediment can also be studied by separating the samples through the use of a graded series of fine-mesh brass sieves. The sediment particles sort out by size (Table 3-5).

Table 3-5 Sediment classes

Sediment class		Diameter/mm
Sand		2.00 - 0.05
	Very coarse	2.00 - 1.00
	Medium	1.00 - 0.10
	Very fine	0.10 - 0.05
Silt		0.05 - 0.002
	Very coarse	0.05 - 0.02
	Medium	0.02 - 0.01
	Very fine	0.01 - 0.002
Clay		<0.002

Sediment that is sand will have distinct grains that are easily seen and felt. Silt will forma cast when moist but will not form a ribbon when moist. Clay is sticky and plastic when wet and forms a ribbon when squeezed. Some sediment samples may have high concentrations of organic matter (muck) indicating slow decomposition rates and low oxygen conditions. Sediments contain minerals (e.g., iron, calcium), which over time, are transformed to limestone, shale, and sandstone.

Exercise: How to identify the texture of the sediment?

1. Take a small handful from your sediment sample.

2. Slowly add little amount of water and mix it very well with the earth sample. Stop adding water as soon as the soil ball straits to stick to your hand. Dry to form into the different shape, till this is no longer possible (Table 3-6).

Table 3-6 Sediment character

Sediment class	Character	Shape
(1) Sand	The soil remains loose and single grained and can only be heaped into a pyramid	
(2) Loamy sand	The soil contains sufficient silt and clay to become somewhat cohesive and can be shaped into a ball that easily falls apart	
(3) Silt loam	The soil can be rolled into a short thick cylinder	
(4) Loam	The soil can be rolled into a cylinder of about 15 cm length	
(5) Clay loam	The soil can be bent into a U	
(6) Light clay	The soil can be bent into a circle that shows cracks	
(7) Heavy clay	The soil can be bent into a circle without showing cracks	

Note: Texture classes (1) to (4) are sandy to silty soils which have generally good infiltration. Texture classes (5) to (7) are clayey soils which have generally poor infiltration.

3.11.5 What organisms are found in the bottom materials?

Samples taken from the Hengshui Lake provide the possibility of observing anaerobic decay. This is especially true in August when biological oxygen demand depletes oxygen in the water above the bottom. Hengshui Lake provides many possibilities for bottom materials study.

Sediments are an important source of nutrients that are released when organic matter decays. When too much organic matter decays, excess oxygen is consumed and eutrophication is stimulated. The materials in samples taken near shore or from the bottom of other bodies of water is basically silt with more or less decayed organic matters above. Oligochaetes (segmented worms related to earthworms), snails, aquatic insects, and shrimps are found there.

3.11.6 What is the connection between sediments and nutrients?

Sediments are an important source of nutrients that are accumulated when they are released from the decay of organic matter. Decay of organic matter consumes oxygen, which can accelerate eutrophication. The "quality" of a lake is shaped by many factors such as its origin and morphology, shoreline development, historical contamination, amount of recreational use it receives, and its overall water quality.

Problems most commonly reported by lake residents are excessive plant growth, algal blooms, and mucky bottom sediments. These can be caused by water quality factors often linked to inadequate management of a lake, which lead to increased lake fertility or productivity. Increased nutrient (nitro-gen and phosphorus) loading leads to degraded water quality and ecosystem health. This loading can be from external (runoff, leaching) and internal sources (sediments).

Cattail is one key emergent plant in Hengshui Lake. It starts to uptake phosphorus from soil since its sprouting in spring. Research on phosphorus uptake of cattail in Hengshui Lake shows that it was highest in June, declined through its growth, and got the lowest in October (Figure 3-10). The phosphorus in cattail would be released back to the soil in sometime of October. For this reason, the total phosphorus removal of cattails will reach its peak in September(Figure 3-11), and the total phosphorus can be removed by harvesting cattails in September.

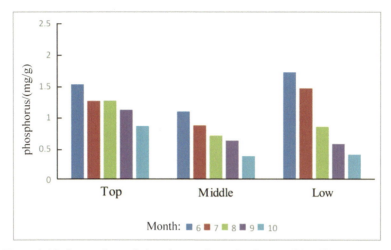

Figure 3-10 Comparison of phosphorus absorption by cattail in different months

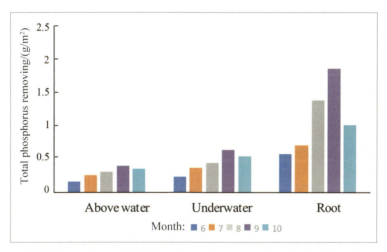

Figure 3-11 Comparison of total phosphorus harvesting in cattail in different months

3.11.7 Frequently asked questions

What is the significance of studying lake sediments?

The comparison of sedimentary facies from different ages is helpful to understand the paleogeography of lake areas. The study of the mineral composition and distribution characteristics of lake sediments, and the identification of the source of sedimentary materials can provide a basis for searching for lacustrine sedimentary mineral deposits. The thickness and nature of the lake can be used to ascertain the age of the lake basin and infer the hydrological and climatic conditions during the formation of these sediments. A large amount of organic matter and a variety of rare elements are accumulated in the sediments, which provide material sources for the formation of various lacustrine deposits.

3.12 Learning unit 12: Plankton sampling

3.12.1 Instructor's guide

Type	Biological testing
Objectives	1. Describe what is a plankton and list types of planktons 2. Catch and identify phytoplankton and zooplankton using plankton nets and field microscopes 3. Identify some typical planktons of the Hengshui Lake
Aids	Plankton net or sampler, microscope, plankton data sheet and drawings of typical planktons

3.12.2 What is plankton?

Plankton, "to drift", is distributed throughout the lake. They are found at all depths and are comprised of both plant (phytoplankton) and animal (zooplankton) forms. Plankton shows a distribution pattern that can be associated with the time of day and seasons.

There are three fundamental sizes of plankton (Table 3-7).

Table 3-7 Type of plankton according to size

Type	Size	Remark
Nannoplankton	5 - 60 μm	Because of their small size, nannoplankton will pass through the pores of a standard sampling net. Special fine mesh nets can be used to capture larger nannoplankton
Microplankton	2 mm	Most planktonic organisms fall into the microplankton or net plankton category
Macroplankton	largest can be several meters long	Visible to the unaided eye

The nets collect the organisms by filtering water through fine meshed cloth. The plankton nets on the vessel are used to collect microplankton.

3.12.3 How is plankton sampled?

The plankton net or sampler is a device that makes it possible to collect both phytoplankton and zooplankton samples (Figure 3-12). For quantitative comparisons of different samples, some nets have a flow meter used to determine the amount of water passing through the collecting net.

Figure 3-12 Plankton net

The plankton net or sampler provides a means of obtaining samples of plankton from various depths so that distribution patterns can be studied. Quantitative determinations can be made by considering the depth of the water column that is sampled. The net can be towed to sample plankton at a single depth (horizontal tow) or lowered down into the water to sample the water column (vertical tow). Another possibility is oblique tows where the net is lowered to a predetermined depth and raised at a constant rate as the vessel moves forward.

3.12.4 What is commonly found in plankton samples?

The base of the food chain in Hengshui Lake is plankton. The phytoplankton is the producer, and they are typically green algae, cyanobacteria (previously known as blue green algae), and diatoms. Cyanobacteria, such as Microcystis, prefer warm water and high nutrients. Some cyanobacteria can fix nitrogen and produce toxins. Crustaceans such as water fleas (Daphnia), representatives of the consumers or zooplankton found in samples.

Use of microscope:

1. Adjust the brightness, from dark to bright, adjust the angle of the mirror.

2. Fix the temporary mount in a proper position on the stage.

3. The low-power objective lens is aligned with the light hole, and the lens barrel is adjusted from top to bottom using the coarse collimating screw. Observe the microscope from the side to prevent the objective lens from contacting the glass slide, damaging the lens or crushing the glass slide.

4. The left eye observes the change in the field of view through the eyepiece, while adjusting the coarse collimation screw. The lens barrel slowly moves up until the field of view is clear (Figure 3-13).

5. If there is no object to be observed in the field of view, you can move the film, the principle is that if you want to go up, go down; and if you want to turn left, turn right.

6. If it is not clear enough, it can be further adjusted with a fine focus screw.

7. If you need to observe under a high magnification objective lens, you can turn the converter to change the objective lens. If the field of view is dark, it can be adjusted by step 1; if it is not clear enough, it can be adjusted by step 6, but not step 4.

8. After use, please adjust the converter so that the empty lens hole is facing the light-passing hole. The reflector is upright, and the lens barrel is adjusted to the lowest point and then installed in the mirror box.

Figure 3-13 Use of microscope

3.12.5 Frequently asked questions

What is lake eutrophication, and what is the dangers of eutrophication?

Lake eutrophication refers to the phenomenon that the lake water receives excessive nitrogen, phosphorus and other nutrient materials, which causes the over-proliferation of algae and other plankton in the water body, the decline of dissolved oxygen in the water body, and the deterioration of water quality and the massive death of fish and other organisms.

The dangers include giving off a fishy smell, decreasing the transparency and dissolved oxygen of water bodies, releasing toxic substances into water bodies, harmful nitrates and nitrites, etc.

Chapter 4 Data recording and analysing of the water science activities

4.1 Data recording

On trips of the floating classroom scheme, a data sheet is prepared listing information about all the parameters that were measured. At the end of a season, these data are entered into an Excel spreadsheet (see Appendices). Information is available on the sampling date, location, latitude, longitude, and depth at each sampling site. Both top and bottom measurements are available for turbidity, conductivity, temperature, pH, and dissolved oxygen. Secchi disk readings, Forel-Ule color scale numbers, listing of benthic organisms, relative plankton density, and sediment types are also in the data set. Alkalinity measurements are listed when they have been taken.

4.2 Data analysing

1. Use the data collected on the vessel to plot the temperature at each sampling site.

2. Compare the graphs, point out similarities and differences, and try to

explain them.

3. Develop a chart that will allow you to show the types and relative abundance of benthic organisms collected from each sampling site.

4. Develop a chart that will allow you to show the types and relative abundance of planktonic organisms for each of the sampling sites.

4.3 Aquatic ecosystem analysing

1. Of all the data collected, list those that show the greatest similarities between the two aquatic ecosystems. List those data that show the greatest differences between the two systems. After reviewing all these data, describe in your own words, why you think these aquatic ecosystems are different.

2. What is the trophic status (eutrophic, mesotrophic, oligotrophic) of the sites in the Hengshui Lake that you sampled?

4.4 Food chain analysing

1. Diagram the planktonic food chain for each of the sampling sites. How are they the same, and how are they different?

2. Diagram a benthic food chain for each of the sampling sites. How are they the same, and how are they different?

Chapter 5 Extension of the water science activities

At the end of each cruise, school pupils summarize their experiences. If conditions permit, sustainability and challenge like climate change should also be highlighted. The follow-up can be also done indoors in the ESD education center or in the schools. A short summary at the end and a longer follow up in the schools is the most appropriate.

5.1 Data maintenance

Data maintenance is a normal extension of water science activities. For instance, a statistics student analyzed a year of data for several locations and determined trends. Classes can use the data for hypothesis testing and to understand how water quality parameters vary by location and season. A variety of statistical measures, charts, and graphs can be generated using the data.

The Monitoring Unit of the NRPB maintains student databases. An online version of data for downloading is planned on the homepage of the biodiversity data base website.

5.2 Questions for experiencers to research

1. What are the two largest lakes of North China? Which has the smallest

surface area?

2. Where do you get the water for your house? Where does your waste water go?

3. When a power plant burns coal, how is the coal soot prevented from escaping the stack? What does escape from the stack? Why is water taken by power plants from the lake? When water is returned to the river or lake, what precautions must be taken? What is made of recycled lime from power plants?

4. How have shoreline developments such as ports, houses, recreational areas, and angling impacted the aquatic ecosystem? What has been done in the area to promote tourism, and how have visitors impacted the area?

5. Describe the bottom sediments from the Hengshui Lake. Why are these sediments like this? What organisms live in these sediments? How are they able to exist at such low oxygen levels?

6. The Yellow River meandered through the HLNR area thousands of years ago, how can you see this today, and why have the former river sites been excellent for brick making?

7. What is biodiversity? What types of organisms live in the water of the Hengshui Lake?

8. What kinds of fish will you find in the Hengshui Lake? Why? How does dissolved oxygen relate to the fish?

9. What is the turbidity in NTUs for pure water? For water from the Hengshui Lake, what time of the year will have the lowest turbidity? Why?

10. What things will cause conductivity readings to be high? What range of conductivity should we expect in the Hengshui Lake?

Appendices

Table 1:

Hydrosphere Data Recording Sheet

Full name: _____ Grade: _____

Instructor: _____

School: _____

Date: _____

(pH paper ☐   pH meter ☐)

Number	Water Depth /cm	Water Temp. /°C				Clarity (Rope Length/m)				Water color				Turbidity (Water Depth/cm)				Conductivity/ (μS·cm⁻¹)	pH				Alkalinity / (mg·L⁻¹)				DO /(mg·L⁻¹)			
		1	2	3	Aver	1	2	3	Aver	1	2	3	Aver	1	2	3	Aver		1	2	3	Aver	1	2	3	Aver	1	2	3	Aver
1																														
2																														
3																														
4																														
5																														
6																														
7																														
8																														
9																														
10																														

Latitude: _____ N   Longitude: _____ E   Time: _____ am/pm

Air Temp.: _____ °C   Wind Direction: _____   Wind Speed: _____ km/h

Table 2:

Benthos and Sediment Data Sheet

Latitude:_____N   Longitude:_____E   Water Depth:_____m   Bottom Temp.:_____°C

Sediment Class:_____   Rope release length:_____m   Sampling Times:_____

Sampler:_____   recorder:_____   Date:_____   Time:_____am/pm

Number: _____

No.	Benthic Organisms	Total Number /ind	Retrieve Number /ind	Notes
Record of Advantage Types				
1				
2				
3				
4				
5				
6				
7				
8				
9				
10				
11				
12				
13				
14				
15				
16				
17				
18				
19				
20				
21				
22				
23				
24				
25				
26				
27				
28				

Table 3:

Plankton Data Sheet

Latitude:_____N   Longitude:_____E   Water Depth:_____m   Water Temp.:_____°C

Sampler:_____   recorder:_____   Date:_____   Time:_____am/pm

Number:_____

Sampling items		Bottle Number	Rope Length /m	Dip Angle /(°　)		Flowmeter		Notes
				Start	End	No.	Revolutions/r	
Trawl Sampling	net							
	net							
	net							
	net							
	net							
	net							
	net							
Water Sampler Sampling							Water extraction volume /cm³	
	Layer							
	Layer							
	Layer							
	Layer							
	Layer							
	Layer							